T0205582

Environmental Footprints and Eco-design of Products and Processes

Series Editor

Subramanian Senthilkannan Muthu, Head of Sustainability - SgT Group and API, Hong Kong, Kowloon, Hong Kong

Indexed by Scopus

This series aims to broadly cover all the aspects related to environmental assessment of products, development of environmental and ecological indicators and eco-design of various products and processes. Below are the areas fall under the aims and scope of this series, but not limited to: Environmental Life Cycle Assessment; Social Life Cycle Assessment; Organizational and Product Carbon Footprints; Ecological, Energy and Water Footprints; Life cycle costing; Environmental and sustainable indicators; Environmental impact assessment methods and tools; Eco-design (sustainable design) aspects and tools; Biodegradation studies; Recycling; Solid waste management; Environmental and social audits; Green Purchasing and tools; Product environmental footprints; Environmental management standards and regulations; Eco-labels; Green Claims and green washing; Assessment of sustainability aspects.

More information about this series at http://www.springer.com/series/13340

Subramanian Senthilkannan Muthu
Editor

Blockchain Technologies for Sustainability

Editor
Subramanian Senthilkannan Muthu
SgT Group and API
Kowloon, Hong Kong

ISSN 2345-7651 ISSN 2345-766X (electronic)
Environmental Footprints and Eco-design of Products and Processes
ISBN 978-981-16-6303-1 ISBN 978-981-16-6301-7 (eBook)
https://doi.org/10.1007/978-981-16-6301-7

This Springer imprint is published by the registered company Springer Nature Singapore Pte Ltd.
The registered company address is: 152 Beach Road, #21-01/04 Gateway East, Singapore 189721, Singapore

Contents

About the Editor

Dr. Subramanian Senthilkannan Muthu currently works for SgT Group as the head of Sustainability and is based out of Hong Kong. He earned his Ph.D. from The Hong Kong Polytechnic University and is a renowned expert in the areas of Environmental Sustainability in Textiles & Clothing Supply Chain, Product Life Cycle Assessment (LCA) and Product Carbon Footprint Assessment (PCF) in various industrial sectors. He has five years of industrial experience in textile manufacturing, research and development and textile testing and over a decade's of experience in life cycle assessment (LCA), carbon and ecological footprints assessment of various consumer products. He has published more than 100 research publications, written numerous chapters and authored/edited over 100 books in the areas of Carbon Footprint, Recycling, Environmental Assessment and Environmental Sustainability.

Blockchain Technology: A Fundamental Overview

Ashraf Jaradat, Omar Ali, and Ahmad AlAhmad

Abstract In November 2008, Satoshi Nakamoto announced the concept of blockchain in a whitepaper about digital payment approach called Bitcoin. Since then, several data-driven domains have identified the potential of using this technology to provide a secure and a trusted platform for transactions processing. In fact, these domains have recognized blockchain as a revolutionized platform of data sharing and data storage. Blockchain technology can be presented as a protocol or software, which offers a secure transfer and storage of exclusive instances of values via the internet and without the need for third-party intermediation. From the functional perspective, blockchain can serve as an open, distributed, and immutable ledger that records transactions in a verifiable and permanent fashion. To understand the revolutionary consideration of blockchain technology, a fundamental overview of the technology is required. Accordingly, this chapter will give an explanation of blockchain concept, its fundamental features, components, categories, and architectures. To emphasize the importance of utilizing this technology in several data-driven domains, a section that addresses the usage of blockchain technology will be presented. Moreover, this chapter will highlight certain aspects related to the future utilization of blockchain technology from Artificial Intelligent (AI) and digital privacy implementations.

Keywords Blockchain · Consensus models · Blockchain architecture · Digital privacy

A. Jaradat · O. Ali (✉) · A. AlAhmad
Department of MIS, Business, and Administration College, American University of the Middle East (AUM), 54200 Egaila, Kuwait
e-mail: Omar.Ali@aum.edu.kw

A. Jaradat
e-mail: Ashraf.Jaradat@aum.edu.kw

A. AlAhmad
e-mail: Ahmad.Alahmad@aum.edu.kw

S. S. Muthu (ed.), *Blockchain Technologies for Sustainability*,
Environmental Footprints and Eco-design of Products and Processes,
https://doi.org/10.1007/978-981-16-6301-7_1

1 Blockchain Fundamental

1.1 Background and Evolution

The digital ledgers for tamper evident and tamper resistant are blockchain and these are being implemented using a distributed procedure, that doesn't require a central repository, or central authority, like the government, company, or bank [6]. Keeping in mind the basic level, the user community is enabled for recording of transactions within the shared ledger of that community, to make sure that the blockchain network operates normally and after publishing the transactions cannot be altered. This blockchain concept, in 2008, was integrated with various kinds of technologies as well as computing concepts so that modern cryptocurrencies could be brought forward. Cryptocurrencies are electronic cash that isn't protected by an authority or central repository but cryptographic mechanisms [94].

In 2009, the bitcoin network was launched and it was then that the technology became widely known. This bitcoin network is one of the first modern cryptocurrencies. For bitcoin and other related systems, the digital data transferred indicates electronic cash being carried out using a distributed system [94].

The users of Bitcoin are able to digitally sign and then transfer their information rights to any other user and a public record is made by the bitcoin blockchain for this transfer. Each network participant is provided the opportunity to verify the transaction validity in an independent manner [42]. A participant distributed group is responsible for maintenance and management, independently, of the bitcoin blockchain. Such a group, integrated with cryptographic instruments, allows the blockchain to avoid any sort of changes to the ledger (transaction forging or blocks identification) at a later point in time. Through this blockchain technology, various kinds of cryptocurrency systems have been developed like the bitcoin and Ethereum [94]. Hence, bitcoin or any other cryptocurrency solutions and blockchain technology are usually considered to be bound. Yet, this technology can be applied to a significant number of applications and several sectors are analyzing its potential [2].

During the late 1980s and early 1990s, this concept of blockchain technology emerged. The Paxos protocol was developed by Leslie Lamport in 1989 and by 1998, a consensus model was presented by this author. This model would form agreements over an outcome within a computer network in the case where the network or computer would be unreliable [50, 94]. During 1991, the electronic ledger used was a signed information chain. This ledger allowed document signing in a manner that indicated that no signed document within this collection was altered [62]. In 2008, the earlier concepts were integrated with electronic cash and referred to as bitcoin. The bitcoin is stated as the peer-to-peer (P2P) electronic cash system [61]. Nakamoto [61] carried out a research which included a blueprint regarding the modern cryptocurrency schemes. There were some modifications and variations subjected. Bitcoin was the first out of various blockchain applications that were developed [94].

Earlier than bitcoin, there were several electronic cash schemes, however, these were unable to attain extensive application [6]. Bitcoin, enabled by blockchain, could be applied in a distributed manner where a specific user was not allowed to control the electronic cash alone and there was no specific failure point present either. Hence, the use of bitcoin was highly promoted [94]. In this system, direct transactions could occur among parties and a third trusted party was not required [94].

New cryptocurrency was also required to be issued in a defined way for users that could publish new blocks and manage the ledger copies. In Bitcoin, these users are referred to as miners. The miners' automated payment helped the system to maintain distributed administration and organization was not required [5]. When the maintenance was based on consensus and blockchain, the self-policing policy was established to make sure the transactions remained valid and the blockchain included the blocks.

The blockchain, within bitcoin, allowed users to remain pseudonymous. It refers to the anonymity of the user, however, the account identifiers do not remain anonymous. Furthermore, the public can view the transactions. Hence, pseudo-anonymity has been offered by bitcoin effectively as the accounts are established without authorization or identification processes [94].

As bitcoin was pseudonymous, it had become necessary to establish procedures that established trust within the environment and it wasn't possible to easily identify the users. Earlier, before blockchain technology was used, an intermediary, which both parties trusted, was used to deliver the trust. After removing these trust intermediaries, the blockchain network helps establish trust through four main features of this technology.

They are:

Ledger—Append only ledger is used by the technology so that the complete transactional history can be presented. As compared to the traditional database, overriding of the values and transactions is not observed in blockchain.

Secure—Cryptographic security is presented by blockchains as it makes sure ledger data is not altered and this data is also attestable [58, 60].

Shared—Various participants share the ledger [13]. Within the blockchain network, such sharing allows for transparency among the node participants [15, 21].

Distributed—It is possible to distribute the blockchain [95]. The blockchain network nodes are scaled so it is highly resilient toward the bad actor attacks [74]. When the nodes number is increased, there is a reduction in the ability of any bad actor to influence the blockchain consensus protocol [74, 95].

The blockchain networks which enable people to create accounts in an anonymous manner and then participate are referred to as permissionless blockchain networks. Such networks help establish trust between the parties who do not have any earlier knowledge regarding one another. Direct transaction can be carried out between organizations and individuals and this would allow for transactions to be quick and low in cost. A blockchain network that controls access in a tight manner is referred to as permissioned blockchain network. Such a network allows limited trust among users and would establish activities that increase trust.

1.2 Blockchain Definition

The blockchain has been defined as a decentralized ledger that manages transaction records on various computers at the same time. For recording transactions, several cryptocurrencies apply blockchain technology. The infrastructure of the bitcoin network is based on blockchain technology [2].

Since the time blockchain has developed, there are various components and analogous terminologies which create confusion regarding the discussion related to technology and how it is implemented. Out of these, there are descriptions associated with the transaction grouping as well as publishing. They are formed into unique data structures referred to as blocks which are cryptographically linked (chained) with each other and spread over the P2P network for the prevention of tampering with earlier published transactions [13, 68]. When the transactions are embedded within the blocks, it requires validation through the nodes (miners) using a consensus model that helps indicate the node given privilege for the next block publication [97]. Blocks also have the ability to support smart contract abilities which are expressive and native using a code collection and the data is presented with the help of transactions that are signed cryptographically over the blockchain network (e.g., Ethereum's smart contracts, Hyper ledger Fabric's chain code) [68, 97].

The pseudonym, Satoshi Nakamoto, stated and implemented the initial blockchain [88]. Nakamoto presented a paper, 'bitcoin in 2008: A Peer-to-Peer Electronic Cash System', in which he stated the blockchain description. It provided the architecture for a payment system that was decentralized and without trust. The blockchain technology is included within the Bitcoin [78], and the cryptographic technology has been used to establish this blockchain [69]. A sequential distributed database is included within this blockchain technology and to save the entire transaction history within a blockchain, a public ledger is used [19, 86, 90, 100]. All allied information and bitcoin transactions are recorded in the bitcoin blockchain [16]. The blockchain is observed to be a public ledger that is distributed and includes records that are immutable and complete of all transactions executed [46]. Even though blockchains and cryptocurrencies are integrated with one another, blockchain technology has gained interest of researchers. It seems to have a revolutionary influence over bitcoin among other things [100] (see Fig. 1).

In this blockchain, the transaction validity and security are carried out using nodes in the P2P network and the blockchain complete record is held [60, 94]. If there is an attack where the blockchain is to be altered fraudulently, then all blockchain copies would be targeted within the entire network. This activity would not be feasible [78]. The blockchain only accepts new transactions with new blocks if verification protocols agreed upon are being followed. When presenting an example for bitcoin, the protocol is referred to as 'proof-of-work', and addition of new transactions can only be done in this way [61, 78].

Blockchain is not only an economic but also a technical alteration [19, 53, 100, 102]. In terms of technical alterations, the blockchain has been considered a new database system form, specifically designed to decentralize environments embedded

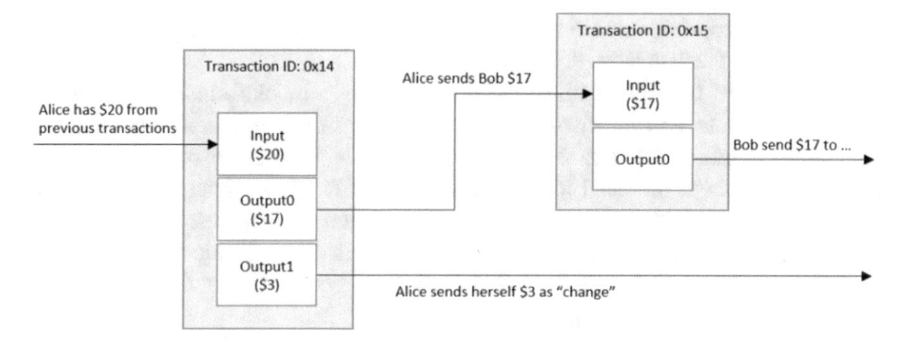

Fig. 1 An example of a cryptocurrency transaction. *Source* Yaga et al. [94]

with lower trust levels. Blockchain, in an innovative sense, extends tools to manage issues that have a transaction record that is dependable and, in the case, where it is not possible to trust the parties, humans or machines [54].

Keeping in mind the mentioned analysis, the blockchain definition established is:

> A distributed database model which relies upon a P2P network is referred to as blockchain. There is a blocks sequence present with transactions that are time-stamped and network community confirmed. The public-key infrastructure (PKI) secures this model. Once an element is part of the blockchain, it cannot be altered. Therefore, the blockchain allows for earlier action recording without disputes [6].

The blockchain technology has various components and relies upon cryptographic primitives along with distribution systems which is why it is quite complex. Yet, simple explanations are present for each component and can be applied as a building block for the complete challenging system. Hence, the informal definition of blockchain is:

> The distributed digital ledgers with cryptographically signed transactions which have been grouped into blocks are referred to as blockchains. There is a cryptographic link of each block with the earlier one (to make sure its tamper evident) after it has been validated and had a consensus decision. It is difficult to modify the old blocks once the new blocks have been included (establishing tamper resistance). Across the ledger copies, in the network, replication of new blocks is done and automatic resolution of conflicts is done using the determined rules [94].

2 Blockchain Components

Blockchain technology seems to be complex technology due to various perspectives and components that need to be associated with it to make blockchain functional and beneficial to the participated stakeholders [45, 94]. However, to simplify the discussion about blockchain technology, this section will individually address the main components of blockchain, such as blockchain transaction, blocks and how they are chained, data store (i.e., ledger), nodes, minors, cryptography primitives, and consensus mechanisms [8, 17, 45, 57, 94].

Transaction: It is the smallest building block of a blockchain system, such as records, contracts, and information, which serves as the purpose of blockchain. It represents an interaction between parties that enable the transfer of value between two parties without the need to trust each other or to have a centralized authority [23]. In the digital currency scenarios, a transaction may represent a transfer of the cryptocurrency between two blockchain network users. While for business-to-business scenarios, a transaction could be a way of recording an activity that occurs on a digital or physical asset [94].

Since a transaction in blockchain is programable, it can hold the applicable data for that program. Even though, the data that encompasses a transaction can differ in every blockchain implementation, the transaction execution mechanism is largely the same [94]. In this execution, a user in a blockchain sends information/ message to the network, where different authentication and validation processes will be carried out at the network. In this form, the sender creates a transaction that may include the sender's address, receiver's public key, the transaction's input, and output values, and the digital signature to prove the authenticity of sent information [8, 23]. Transaction inputs often signify a list of the digital assets to be transferred, while transaction outputs signify the recipients' accounts and amounts of the transferred digital assets. Figure 1 illustrates the execution of a cryptocurrency transaction.

Node: Basically, any entity that connects to the blockchain network, is called a node or a miner, where each node can have an independent copy of the whole blockchain ledger [23, 94]. The subset of nodes in the blockchain network that are equipped with special software to validate the transactions or to solve a difficult math problem to be able to find a new block is called miners [63]. By using a computer's processing power, the mining process is mandatory for recording transactions on the blockchain network. Additionally, as nodes are used to verify all the blockchain rules, then these nodes are called full nodes. Full nodes group the transactions into blocks to determine the validity of the transaction [23].

Block: It is one of the main blockchain network components as a blockchain is represented in sort of a sequence of blocks [57]. Each block stores a complete list of transactions records, similar to the public ledger. The first block in the sequence is called genesis block. This block does not have a parent block. The blocks are linked to each other using a reference hash for the previous block. This block is known as the parent block, such that each block has only one parent [57, 101]. Figure 2 shows blockchain structure in a sort of continuous sequence of blocks.

An individual block consists of block header and block body as shown in Fig. 3. The block header includes a set of metadata information such as block validation rules called Block version, timestamp, Merkle-tree root hash that represents the hash value of all the transactions in the block, nBits target threshold of a valid block hash, Nonce (i.e., 4-Byte field), and parent block hash that points to the previous block in the chain. The Block body is composed of the actual transaction data and the transaction counter. It is the part of the block that affects the upper limit of a possible transaction and its actual time [57, 64, 75, 101].

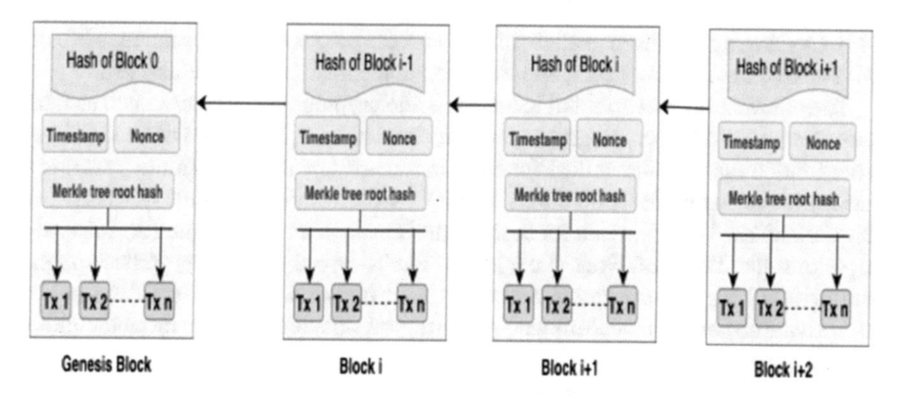

Fig. 2 Blockchain structure in a sort of continuous chain of blocks. *Source* Monrat et al. [57]

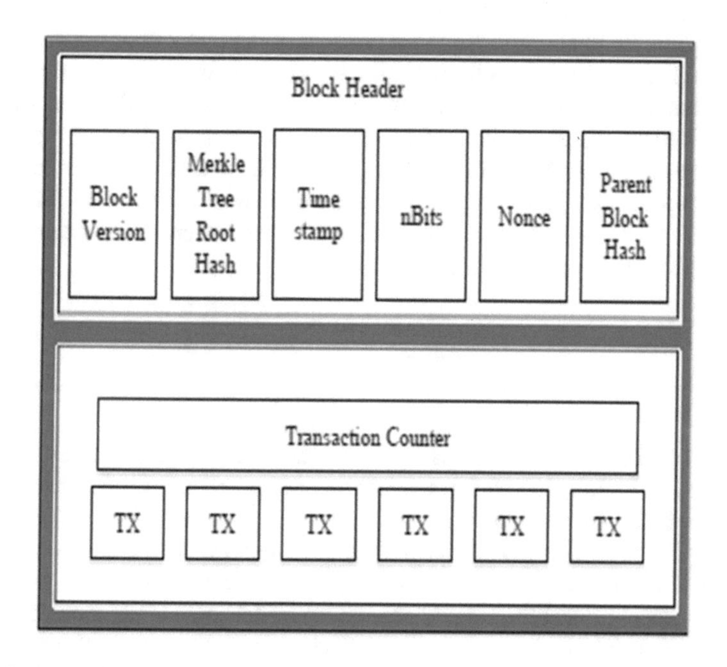

Fig. 3 Structure of a blockchain block. *Source* Niranjanamurthy et al. [64]

Ledger: It is data store in a blockchain that stores immutable, sequenced records in blocks using a consensus algorithm [8]. There is generally one ledger per channel. Each node maintains a copy of the ledger for each channel that it is a member of. The shared ledger encodes the entire transaction history for each channel as well as includes query capabilities for efficient processing [8, 63].

Cryptography: it is the most important component in a blockchain system as it relies extensively on cryptography throughout various steps within system operations

[47]. Cryptography is used to allow relevant access, to store data in immutable blocks with a fixed consecutive order, and to establish identity and authenticity [8].

To achieve a fully distributed ledger, specific cryptography aspects, such as hash functions, public-key cryptography, and digital signature, have to be utilized [47]. A hash function is a mathematical function that maps any data as input to generate a hash value as an output for all the transactions that are under a specific threshold [8]. Blockchain used this function for hashing the transaction message and for consensus algorithm like Proof-of-Work (PoW) [47]. Public-key-cryptography utilizes public and private keys for encrypting and decrypting transactions messages. The public and private key pairs are also utilized by the digital signature to prove the authenticity of the data [8].

Consensus: It is another blockchain's key component is related to the way data entries are being accepted onto the distributed ledger. Blockchain technology uses consensus algorithms to maintain a single history of blocks by synchronizing the records within the chain of blocks to ensure that each block does not contain any contradicted or invalid transactions [39]. Several consensus algorithms have been proposed and used in the blockchain system implementation [25]. Basically, those proposed consensus algorithms can fall under two broad categories, i.e., proof-based and vote-based. Proof-of-Work (PoW) or Proof-of-Stake (PoS) is a common proof-based consensus algorithms that are popularly used with public blockchain, while Byzantine, which represents an example of vote-based consensus algorithms, is used widely in private blockchain [25].

3 Blockchain Features

When the blockchain technology is to be assessed, the essential features to be kept in mind are audibility, anonymity, persistency, decentralization, and trust [78, 86]. These have been discussed further.

Trust: The blockchain essential feature is embedded within the decentralized approach [78]. Transactions are not secured by a third party or assets as a proof-of-work protocol protect the network, hence, no middle man is required for trusting of transaction recording or verifying [19]. There is an open source of the complete blockchain and anyone can view it, hence, no backdoors are built within the system [5]. Individuals make sure their capital remains safe by being in control and as the bank themselves. The traditional banking system, however, requires that the bank controls the money of the customers [43]. Trust in blockchain technology is explained through terms like public and shared interaction [7, 13, 15, 83], peer verification of transactions [32, 34], low friction in extending information [19, 83], and security using cryptography [7, 93, 102].

Decentralization: It is another key aspect of blockchain technology. Using this, immutability and censorship resistance is extracted [78]. Apart from its features, a platform is extended in which there is no need for a third party to protect the assets [86]. The centralized ledger cannot be infiltrated by a hacker or government due to the blockchain's decentralized and distributed nature [6]. Complex mathematical problems can be resolved through proof-of-work by using computing power. Furthermore, decentralized nodes, in thousands and hundreds, are harmonized using the consensus system that is proof-of-work. Therefore, asset safety is assured and prevention is present for money supply arbitrary dilution [78]. There are some terms that refer to blockchain technology decentralization, like, participants' pseudonymity [34, 102] automation potential application [36, 93], data redundancy [13, 43], and 'versatility' development and participation of peers [100].

Persistency: Quick validation of transactions and honest minors wouldn't admit transactions that are invalid [26, 102]. Transactions cannot be rolled back or deleted once the blockchain includes those [86]. Immediate discovery is made for invalid transactions within a blockchain [78].

Anonymity: Once an address is generated, users may interact with the blockchain without revealing their identity [27, 51]. Intrinsic constraints are present which is why the blockchain may not guarantee privacy a hundred percent [26, 34].

Auditability: The Unspent Transaction Output (UTXO) model is used by blockchain for storing information regarding user balances [61, 101]. All transactions are referred to an earlier transaction that wasn't spent [101]. Once the blockchain records the present transaction, the unspent transactions are changed to spend. Thus, transactions can be verified and tracked easily [13, 43].

4 Blockchain Architecture

The technology of blockchain works on the concept of decentralized databased where each database can exist in multiple computers and each copy of these databases is identical [75]. Blockchain architecture illustrates the connection of nodes that are existed on a network for validation and data transaction purposes [44]. It also refers to a distributed system, data structure, and/or network of blocks that are ordered in a list form [84]. Accordingly, a blockchain architecture can mainly be divided into three layers, i.e., Application layer, Decentralized ledger layer, and Peer-to-Peer (P2P) network layer. Figure 4 shows the blockchain architecture based on these three layers.

The application layer contains the applications logic of the blockchain. It provides a readable interface where users can keep track of their transactions. For instance, Bitcoin wallet software creates and store private and public keys that allow users to control the unspent amount of the bitcoin [75]. This layer is collectively controlled by the participants who deploy the application code onto the blockchain network when it is operational. The decentralized ledger is the intermediary layer in blockchain architecture. This layer reinforces a consistent and tamper-proof global ledger for a blockchain architecture. Transactions in this layer are grouped into crypto-graphical and linked blocks. The mining process in a distributed ledger is used to group transactions into blocks, which is added to the end of the recent blockchain. Transactions in this layer can be presented as an exchange of tokens between two participants where

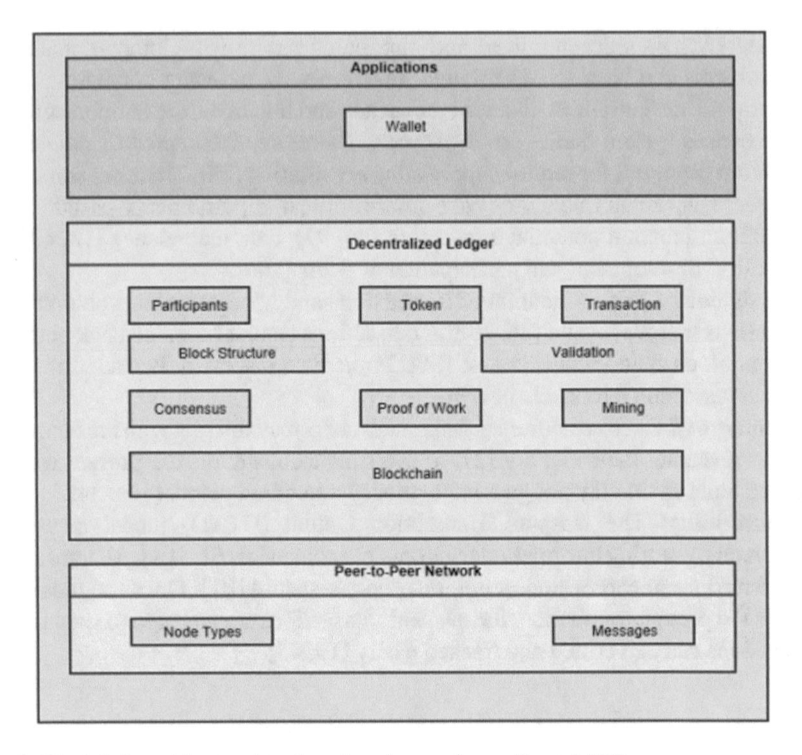

Fig. 4 Blockchain architecture based on three layers. *Source* Sarmah [75]

every transaction goes through a validation process before it is considered legitimate. The bottom layer in the blockchain architecture is the P2P Network where different types of nodes play different roles and multiples messages are exchanged with a decentralized ledger [75, 101].

5 How Does Blockchain Work?

A blockchain records data in a sequential archive [96]. This structure proposal was initiated by [37] for time-stamping intellectual property rights before it can be copied by others. In their model, Haber and Stornetta assured the authenticity of each timestamp using hash functions that transform data into hexadecimal code of fixed length to prevent the inversion to recover the original inputs. The authors also proposed the continuous transformation of each entry in their sequence into hash code that would be combined with the data for the next entry. By this form of records archiving, the time of digital documents creation would be authenticated, which will help the prevention of forge [96].

Consequently, the working nature of blockchain starts with someone requesting a transaction. Then this requested transaction is broadcasted to a P2P network that consists of nodes of digital devices. After that, the network of nodes will validate the transaction and the user status based on specific consensus algorithms. A verified transaction can include patient record, cryptocurrency, digital contract, or any other data transaction. Once the transaction is verified, it will be linked to other transactions to create a new block of data for the distributed ledger. This new block will be appended to the existing blockchain. The appointment will be observed by presenting the new block in the blockchain in a way that is permanent and unchangeable. As the transaction is added to the chain, then we finally can state that the transaction is completed [64, 75]. Figure 5 depicts the working nature of a blockchain to transfer money between two individuals.

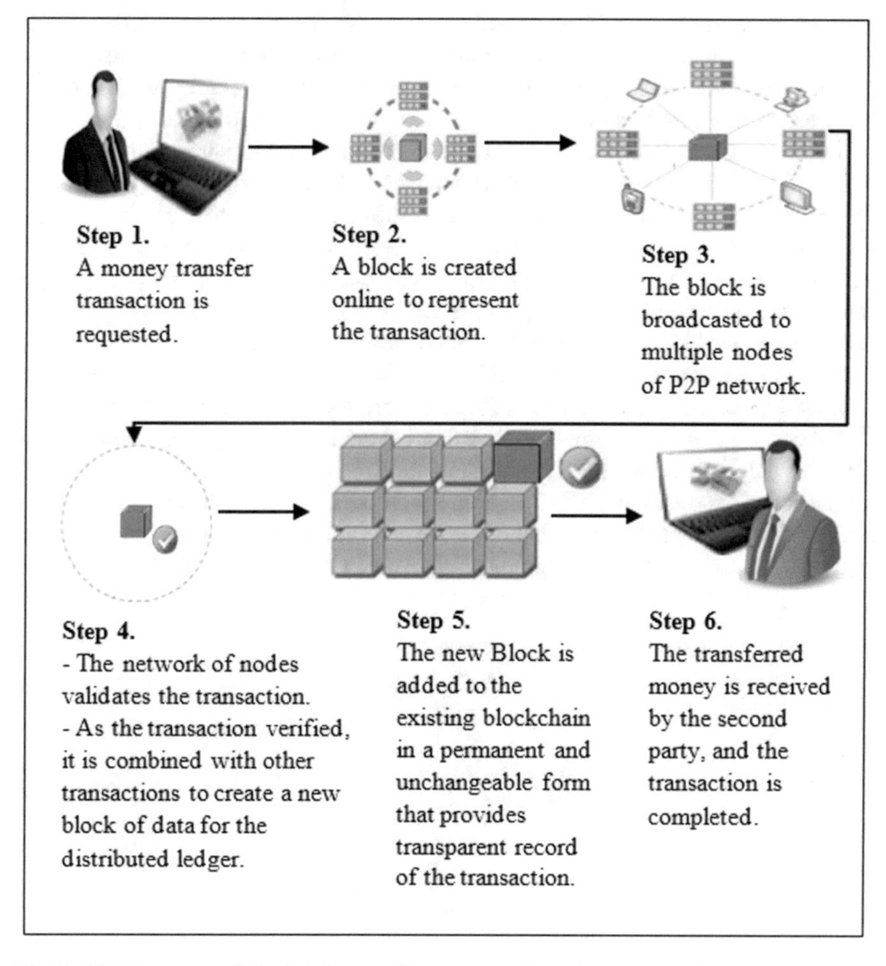

Fig. 5 Working nature of blockchain to perform a transaction of money transfer

6 Blockchain Categorization

Blockchain has been evolved greatly in the last few years, where different researchers proposed categorization schemes based on varied perspectives such as network management, permission models, decentralization degree and immutability, efficiency, and consensus determinations. For instance [75], introduced in his research paper ten types of blockchain such as public, private; semi-private; fully private of proprietary, sidechains; permissioned ledger; distributed ledger; shared ledger; tokenized; and tokenless blockchains. While other research papers as [94, 99] presented blockchain in sort of two major categories, i.e., permissionless (public) as anyone can publish a new block, and permissioned (private) as only particular participants can publish blocks. However, majority of blockchain research works such as [7, 44, 47, 57] identified and differentiate between three different blockchain network categories: Public (Unpermissioned), Private (permissioned), and Federated (Consortium) blockchains. The following sections will explain the main characteristics and differences of each category. At the end of this section, a comparison between these categories based on different blockchain properties perspectives is outlined.

Public blockchain: It is recognized as permissionless ledger such as Bitcoin, Litecoin, and Ethereum, where interconnected nodes on the network are easily accessible by anyone via the internet [44, 47]. Figure 6 shows the public blockchain network. Public blockchain consists of an open and publicly accessible network, such that any network member can validate a transaction and participate in the approval process through the utilization of such consensus algorithms as PoW) or PoS [47]. Participants on a public blockchain network do not require to know or trust each other to have a secure transaction transfer. Within this trustless platform, transaction validators get rewarded with economic incentives for their honest behavior. Accomplishing individual control over the entire network is hard to achieve, hence, public blockchains offer very high protection level against unauthorized data manipulation [47]. In fact, a transaction on the blockchain ledger is synchronized at all nodes in the network, such that anyone who has a computer device and internet connection will be able to register as a node and offer a complete blockchain record [48, 75]. The repetition

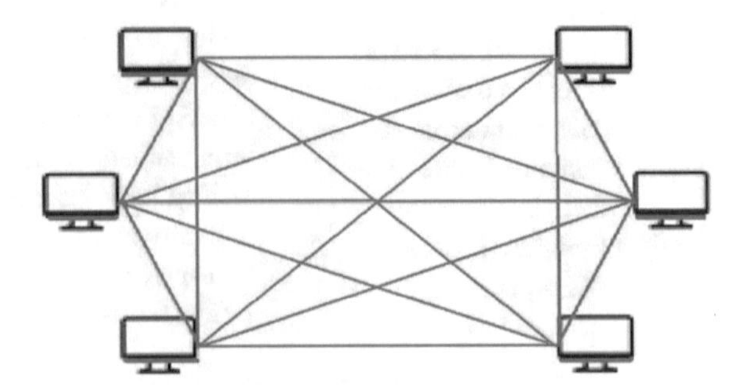

Fig. 6 Public blockchain network. *Source* Niranjanamurthy et al. [64]

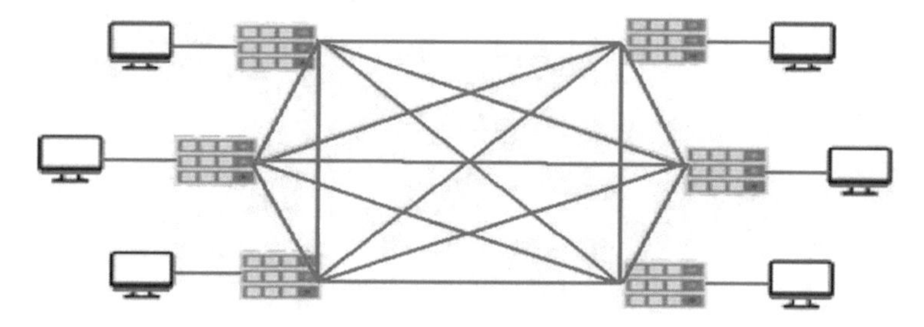

Fig. 7 Private Blockchain network. *Source* Niranjanamurthy et al. [64]

of synchronization at each node in the public blockchains network assure the security of transaction processes. Moreover, due to the decentralization nature, a public blockchain will facilitate true editing resistance.

However, this blockchain category suffers from performance constraints. It is subject to of poor scalability with gradual and inefficient processing rates of transactions validations. Moreover, huge electrical consumption and power supply are required to validate each transaction or when each node is added to the network [44, 47].

Private Blockchain: This category of restricted blockchain is permissioned and centralized to one governing organization [75]. Private Blockchains are based on closed networks, where participants can join the network based on special permissions to read, write, and validate transactions. Figure 7 shows the private blockchain. Transactions are validated internally by a closed group of preselected authorities, which have full control of all stored data. A high level of efficiency will be offered in the verification and validation of transactions in private blockchain [44, 47].

This category of permissioned blockchain comes with low level of decentralization, where all shared data is hardly distributed [47]. Therefore, data structure in private blockchain is not immutable and it loses its ability to provide effective data protection against manipulation [75]. It also enables single point of failures, hence, availability and failure tolerance of the stored data is decreased [57]. This category can be represented as a trust transformer such that trust is based on an algorithm more than an authority [48]. Accordingly, the participants on private blockchain network should trust a centralized group of validators. This concern is reflected in the asset's transfer and on the data storage under a specific account. Since all validators know and trust each other, the need of economic incentives is not required [47]. As a malicious participant being identified in the network, the validators will directly exclude him by their given superordinate authorities. System-wide consensus recognition is trivial due to the limited decentralization degree. Private Blockchains are highly scalable and they achieve a high rate of transaction processing. For instance, Monax, HyperLedger with Sawtooth, private Ethereum, and Quorum represent few examples of private and permissioned blockchain [48, 57]. Unlike public blockchain, private blockchain does not provide a decentralized system for secured database, which has been considered as one of the main limitations of private blockchain [44].

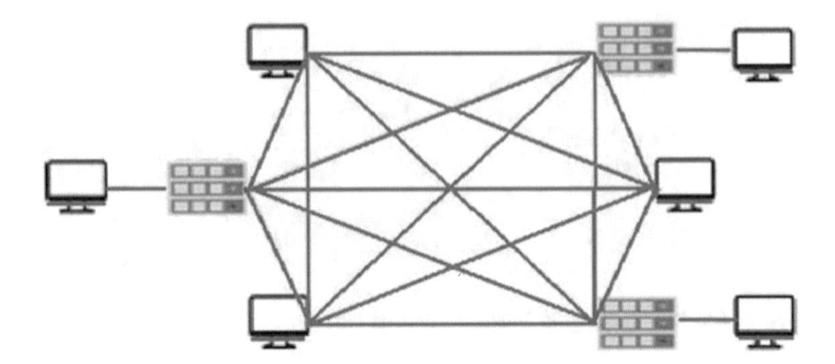

Fig. 8 Federated blockchain network. *Source* Niranjanamurthy et al. [64]

Federated Blockchain: The last category represents the tradeoff between the former public and private blockchain categories, where some parts of the blockchain are private and controlled by a group of individuals or organizations while the rest is open to anyone to participate [57, 75]. Thus, federated blockchain is a hybrid solution that enables a semi open and publicly accessible network that can be presented as partially decentralized [44]. R3 Corda, Ripple (XRP), Stellar, and B3i are blockchain examples related to the federated category [47, 48]. The federated blockhain networks are presented in Fig. 8. Data transactions are mainly transparent as they could be read and written by any network participant. Yet, the validity authorization of propagated transactions is a preselected and/or subsequent privilege is given to a group of validators that forms a consortium [47, 48]. Every member of the consortium will be selected carefully where exclusion can be applied for any incorrect behavior.

Similar to private blockchain, all members in the federated blockchain trust and know each other, and no economic incentive can is observed since no rewards are given for any correct behavior [47]. For this blockchain category, consensus can be established easily as long as no malicious members exist in the network [47]. As result, the complexity of finding consensus is relatively low, and hence federated blockchains are easily scalable and the transaction processing rates are quite high. Moreover, the decentralization and distribution level of all transaction data could be high as well. This depends on the size of the consortium and the architecture of the blockchain system. Accordingly, federated blockchain might offer an immutable data structure and provide effective security protection against data manipulation [47]. Federated blockchains are also a controversial topic as the private ones. For instance, the ability of misusing the power supply is still possible in this category, even though the possibility of misusing power supply has been lowered by distrusting it between public participants through the basics of a transparent consortium [44, 47].

Based on the above discussion, a summarized comparison between these three former categories is outlined in Table 1. This comparison is based on certain blockchain properties and perspectives, such as access permission level, read/ write permission, consensus determination, efficiency, and decentralization degrees.

Table 1 Comparison among the three blockchain categories

Blockchain Properties perspectives	Blockchain categories		
	Public	Private	Federated
Access nature	Publicly open	Closed	Semi open
Validators	Any participant	Closed group	Consortium
Consensus determination/	All minors	Selected set of nodes	Within one group or organization
Read/write permission	Permissionless-Public	Permissioned- could be public or restricted	Permissioned-could be public or restricted
Decentralization	High	Low	Partial
Immutability level	Almost impossible to manipulate/tamper	Could be manipulated/tampered	Controlled and could be manipulated/tampered
Transparency	Low	High	High
Efficiency	Low	High	High
Scalability	Low	High	High
Latency	High	Low	Low
Trust	Low	High	Medium/High
Processing Rate	Low	High	High
Energy Consumption	High	Low	Low

7 Blockchain Regulations and Organization Incorporation

Blockchain has incorporated certain advancements in using many technologies that used to manage critical resources. Those changes made it critically required to apply various amendments to the governments' regulations and organization incorporation. It requires defining blockchain-related processes and standardizing the terminology used. Therefore, these regulations need to be modernized to meet Blockchain decentralization, anonymity, and smart contracts [82].

As mentioned earlier blockchain is one of the most significant developments in information technology toward security and trust. Rodrigues [71] noted that blockchain trust will not be effective without the well-formed supporting governance including regulation including both laws and policies. The uniqueness of blockchain needs to be attained by institutions, doing so needs them to work side by side with blockchain makers, designers, and users as this technology are expanding to most if not all industries and it is not limited anymore to entrepreneurial ventures [92].

On the government level, the regulations must be modernized to be able to consider the digital proofs used in blockchain. For example, this technology can proofs accountability technically, but this proof is not yet considered as legal evidence in most countries [3]. The distribution environment that blockchain use formed another legal challenge that needs special attention, especially when deciding which county law should be applied if parties participated in the blockchain are located in multiple countries [85].

This first requires an assessment of the law, regulations, and policy and to determine how regulators are to legally define blockchain before providing any legal form of use. Bringing blockchain to the courts requires a deep knowledge of this technology's concepts, architecture, and implantation [89]. In other words, judges, lawyers, and practitioners aside from regulators need a tremendous amount of study to be able to confirm that their practices are legal and align with the regulations.

Studies have shown the vast impact of blockchain technology on the business environment, strategies, operations, and supply chains, besides other relevant areas [11]. Normally business adapts to new technologies earlier than government, having said that, the organization incorporation needs to adapt faster in order to explore more of blockchain possibilities toward replacing or enhancing previously used technologies [72].

The business environment has been altered by blockchain wherever it is was implemented. As a result, businesses strive to establish the social and economic legitimacy of blockchain and its applications [100]. This escalates the importance of understanding the guidelines for adopting blockchain through the business environment. Doing so should be done while keeping an eye on the fact that, yes blockchain enhances security, but it does not mean that it is immune against all attacks [93].

In the same context [77], discussed the fact that blockchain's goals in business are to capture the external business environment as well as the internal aspects of an organization. For instance, most of the business environment processes, such as business network identification, network analysis, trust, product offering, product bundling, network context, and the external business environment, are supported by blockchain.

Advances in blockchain are affecting business strategies by controlling where, when, and why any business stands in blockchain networks [30]. For example, blockchain technology can enable the business to provide fully automated smart contracts between organizations. Such implementations require many changes to adjust internal and external processes to adapt to the new modifications, or else you will be left behind.

8 Blockchain Usage

Blockchain nowadays is considered as one of the newest technologies evolve and its usage is spreading and continue to grow exponentially [1]. Liu [55] in her report to Statista has shown that the global market for blockchain solutions is expected to reach 6.6 billion dollars in 2021. According to the Liu [55] report, the projections, spending on blockchain solutions would increase in the next years, reaching about 19 billion dollars by 2024.

Blockchain is used within a wide range of applications in various fields, such as finance and banking, business, education, health care, government. Usage of blockchain in certain industries is shown in Fig. 9. Blockchain seems to have reached all industries and created many opportunities to enhance human life. Among the industries that took the usage leadership of blockchain in finance and banking which have seen it as a technology that cannot be avoided [5].

At first, blockchain was used in cryptocurrencies like Bitcoin where two or more parties, individuals or businesses, are able to make agreements and transactions without any intermediaries to execute clearing, record-keeping, settling, and contracting [22, 87]. The largest individual blockchain technology use case in 2021 was cross-border payments and settlements [55].

For finance and banking, blockchain provided opened the door for many opportunities that can improve the efficiency and effectiveness of their internal processes and the services they provide to their customers. The main benefit of this technology is decentralization which enables trading in real-time. Additionally, blockchain enabled the creation of a trusted and secured system for these financial institutions [98].

In March 2020, Credit Suisse collaborated with Paxos, the first regulated blockchain infrastructure company, to use blockchain technology to settle US stock trades [65]. Meanwhile, JPMorgan Chase has launched the JPM Coin, a cryptocurrency that will be used to simplify transactions between institutional accounts. Other financial institutions, such as Goldman Sachs and Citigroup, have also ventured into blockchain [20].

After the success in finance and banking, blockchain became common to be used in certain business processes, i.e., market exchange, resources management, supply chain, and rights management [91]. The fact that blockchain is a major game-changer in the business has led to the use of this technology in most business domains. Blockchain has supported the business in seeking the direction of revolutionizing existing procedures to enhance efficiency and effectiveness [59].

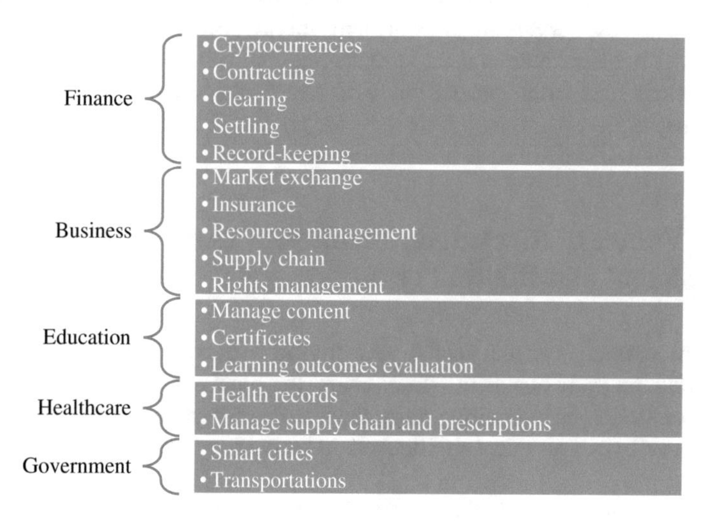

Fig. 9 Blockchain usage in different industries

In education, by nature, any academic credentials need to be recognized universally. In all education levels, most educational institutions are still using the manual process in managing content and verifying academic credentials [18]. The use of blockchain technology in education could simplify managing the contents and enhance verification procedures' effectiveness and efficiency which in turn will be reducing the number of fraudulent claims for unearned credits.

Due to the needs, we discussed above, Blockchain was implemented to manage content [81], learning outcomes evaluation [24], and certificates [35]. For instance, the University of Nicosia used Blockchain technology to manage certificates [80]. Also, Sony Global Education used Blockchain technology to build a platform that certifies degree information in terms of academic proficiency and progress records [41].

The inability of healthcare organizations to safely transfer data across platforms and manage prescriptions report safety and anxiety concerns [56]. More accurate diagnoses and prescriptions, more effective treatments, and more cost-efficient care could all result from improved data sharing across health care providers. The use of blockchain technology enabled the health care institution to transfer data more securely with a high level of productivity and usefulness.

For instance, blockchain is used in the health care sector to manage records [9] and manage prescriptions which in return also help in controlling the drugs [48]. Blockchain is securely used in the creation, storage, and management of the patients' medical and health-related data [49]. Furthermore, it is used in the drugs domain to manage health supply chain management, and certify prescriptions to detect prescription drug fraud [31].

While in the government sector it was mainly used in the smart cities network [79] and public services control and monitoring [83]. This will help in maintain the integration and efficient resources utilization between the elements of humans (who will receive and use service and network), machines, government, and organizations [14]. Likewise, blockchain is used in managing and controlling securely real-time transportations data from sensors, onboard and roadside units, GPS, and driver's smartphones [99].

9 The Future of Blockchain Technology: Artificial Intelligent and Digital Privacy

With cryptocurrency, virtual banks, self-driving cars, and assistants that can do routine tasks on your behalf in casual chats, the blockchain, and AI have started shaping the future. It is obvious these days that most of the invocations that are changing our lives are linked to Blockchain technology and artificial intelligence [29].

AI when adapted to Blockchain improved privacy, trust, reliability, security, management, transparency, and data. Users now can trust their privacy control, and where and how data is used. This privacy control will lead to more applications toward humanity, specifically in AI [76]. Multi-agent, machine learning, analytics, decision-making are examples of future directions in the blockchain with AI integration [73].

AI is a mystery for users who do not dig deep into this complicated science. This fact represents reliability issues for AI technologies. The reliability while lowering the cost of data when using AI will be enhanced by blockchain's transparency. Furthermore, blockchain when integrated with AI can improve the security, trust, and management of data and algorithms in AI applications [73].

Blockchain, on the other hand, can improve AI effectiveness by allowing for secure data sharing, which allows for more data, which leads to more training data and more powerful AI models [28]. Additionally, merging AI and smart contracts may reduce risk scenarios as smart contracts are programmed to do specific actions when specified criteria are met [10].

As an example of integrating blockchain with AI, machine learning which is one of many dimensions in AI is has benefited a lot from using blockchain technology. Harris and Waggoner [38], proposed a model to overcome the centralization issues that machine learning suffers by using blockchain decentralized a dataset and by using smart contracts to host a model that is updated continuously. This integration helps in sharing resources, versioning, allow for smooth evolving, and support trust.

More implantations that integrate blockchain with AI are yet to come. Entrepreneurs are trying to invest in the hidden opportunities in computer vision, context-aware computing, machine learning, and natural language processing [66]. Furthermore, AI with blockchain can also be used in the future in consulting, support and maintenance, and system integration and deployment (Cloud and On-Premises).

Both of these technologies, blockchain and AI, can drive digitalization across almost all economic sectors. By the end of 2025, the global blockchain AI market is estimated to have grown from USD 129.46 million in 2020 to USD 651.56 million [70]. This represents a great opportunity for all practitioners and researchers in both industries AI and blockchain which can help to shape the technology future.

Blockchain openness, transparency, and decentralization made privacy one of the main preservations of users. This technology supports all the privacy dimensions, i.e., privacy of transactions, machines, networks, and participants using access control management that does not require trust in a third party [102]. This privacy protection was enabled since the distributed data that is stored and travel inside the blockchain network is cryptographically protected [40].

For instance [52], proposed to use blockchain technology to provide to enhance the privacy and availability of provenance data in cloud computing. Furthermore, blockchain was used to solve the privacy issues problems faced by privacy protection intelligent parking system based [4], the secret share voting system [12], and transparent voting platform [33].

However blockchain mitigated many privacy issues as mentioned above but still, just like any newborn technology, blockchain has certain challenges in terms of security and specifically in privacy. For instance, among the main concerns that were discussed by Peng [67] are: (i) The extensive computation required in the blockchain which may lead to inefficient privacy schemes (ii) The smart contract along with its data protection (iii) accountability which requires in certain cases to reveal users identity.

References

1. Abou Jaoude J, Saade RG (2019) Blockchain applications—usage in different domains. IEEE Access 7:45360–45381
2. Abramova S, Böhme R (2016) Perceived benefit and risk as multidimensional determinants of bitcoin use: a quantitative exploratory study. In: The 37th international conference on information systems, pp 1–20
3. Al-Jaroodi J, Mohamed N (2019) Blockchain in industries: a survey. IEEE Access 7:36500–36515
4. Al Amiri W, Baza M, Banawan K, Mahmoud M, Alasmary W, Akkaya K (2019) Privacy-preserving smart parking system using blockchain and private information retrieval. In: 2019 international conference on smart applications, communications and networking (SmartNets). IEEE, pp 1–6
5. Ali O, Ally M, Clutterbuck P, Dwivedi Y (2020) The state of play of blockchain technology in financial services sector: a systematic review. Int J Inf Manag. https://doi.org/10.1016/j.ijinfomgt.2020.102199
6. Ali T, Alzahrani A, Jan S, Siddiqui MS, Nadeem A, Alghamdi T (2019) A comparative analysis of blockchain architecture and its applications: problems and recommendations. IEEE Access 7:176838–176869
7. Ali O, Jaradat A, Kulakli A, Abuhalimeh A (2021) A comparative study: blockchain technology utilization benefits, challenges and functionalities. IEEE Access 9:12730–12749
8. Alketbi A, Nasir Q, Talib MA (2018) Blockchain for government services—use cases, security benefits and challenges. In: Proceedings of the 15th learning and technology conference, Jeddah, pp 112–119
9. Angraal S, Krumholz HM, Schulz WL (2017) Blockchain technology: applications in health care. Circ: Cardiovasc Qual Outcomes 10(9):e003800
10. Arumugam SS, Umashankar V, Narendra NC, Badrinath R, Mujumdar AP, Holler J, Hernandez A (2018) IoT enabled smart logistics using smart contracts. In: 2018 8th international conference on logistics, informatics and service sciences (LISS). IEEE, pp 1–6
11. Bai CA, Cordeiro J, Sarkis J (2020) Blockchain technology: business, strategy, the environment, and sustainability. Bus Strategy Environ 29(1):321–322
12. Bartolucci S, Bernat P, Joseph D (2018) SHARVOT: secret SHARe-based VOTing on the blockchain. In: Proceedings of the 1st international workshop on emerging trends in software engineering for blockchain, pp 30–34
13. Beck R, Czepluch JS, Lollike N, Malone S (2016) Blockchain-The gateway to trust-free cryptographic transactions. In: The 24th European conference on information systems (ECIS), pp. 1–14
14. Biswas K, Muthukkumarasamy V (2016) Securing smart cities using blockchain technology. In: 2016 IEEE 18th international conference on high performance computing and communications; IEEE 14th international conference on smart city; IEEE 2nd international conference on data science and systems (HPCC/SmartCity/DSS). IEEE, pp 1392–1393
15. Bonneau J, Miller A, Clark J, Narayanan A, Kroll JA, Felten EW (2015) Research perspectives and challenges for bitcoin and cryptocurrencies. In: IEEE symposium on security and privacy, pp. 104–121
16. Bradbury D (2015) In blocks [security bitcoin]. Eng Technol 10(2):68–71. https://doi.org/10.1049/et.2015.0208

17. Bruyn AS (2017) Blockchain an introduction. University Amsterdam, 26. https://beta.vu.nl/nl/Images/werkstuk-bruyn_tcm235-862258.pdf. Accessed 19 July 2021
18. Bsalian R, Acharya S, Aithal PS (2015) Information technology innovations in office management-a case study. Int J Res Dev Technol Manag Sci (Kailash) 21(6)
19. Böhme R, Christin N, Edelman B, Moore T (2015) Bitcoin: economics, technology, and governance. J Econ Perspect 29(2):213–238
20. CBINSIGHTS (2020) Banking is only the beginning: 58 big industries blockchain could transform. https://www.cbinsights.com/research/industries-disrupted-blockchain
21. Cai Y, Zhu D (2016) Fraud detections for online businesses: a perspective from blockchain technology. Financ Innov 2(20):1–10. https://doi.org/10.1186/s40854-016-0039-4
22. Casey M, Crane J, Gensler G, Johnson S, Narula N (2018) The impact of blockchain technology on finance: a catalyst for change
23. Casino F, Dasaklis TK, Patsakis C (2019) A systematic literature review of blockchain-based applications: current status, classification and open issues Telemat Inform 36:55–81
24. Chen G, Xu B, Lu M, Chen NS (2018) Exploring blockchain technology and its potential applications for education. Smart Learn Environ 5(1):1–10
25. Chukwu E, Garg L (2020) A systematic review of blockchain in healthcare: frameworks, prototypes, and implementations. IEEE Access 8:21196–21214
26. Conti M, Sandeep Kumar E, Lal C, Ruj S (2018) A survey on security and privacy issues of bitcoin. IEEE Commun Surveys Tutor 20(4):3416–3452
27. Dasgupta D, Shrein JM, Gupta KD (2019) A survey of blockchain from security perspective. J Bank Financ Technol 3(1):1–17
28. Dillenberger DN, Novotny P, Zhang Q, Jayachandran P, Gupta H, Hans S, Sarpatwar K (2019) Blockchain analytics and artificial intelligence. IBM J Res Dev 63(2/3):5–1
29. Dinh TN, Thai MT (2018) Ai and blockchain: a disruptive integration. Computer 51(9):48–53
30. Durach CF, Blesik T, von Düring M, Bick M (2021) Blockchain applications in supply chain transactions. J Bus Logist 42(1):7–24
31. Engelhardt MA (2017) Hitching healthcare to the chain: an introduction to blockchain technology in the healthcare sector. Technol Innov Manag Rev 7(10):1–10
32. Eyal I, Gencer AE, Sirer EG, van Renesse R (2016) Bitcoin-NG: ascalable blockchain protocol. In: The 13th USENIX symposium networked systems design and implementation, pp 45–59
33. Faour N (2018) Transparent voting platform based on permissioned blockchain. arXiv:1802.10134
34. Garman C, Green M, Miers I (2014) Decentralized anonymous credentials. In: Network and distributed system security symposium, pp 23–26
35. Gräther W, Kolvenbach S, Ruland R, Schütte J, Torres C, Wendland F (2018) Blockchain for education: lifelong learning passport. In: Proceedings of 1st ERCIM blockchain workshop 2018. European society for socially embedded technologies (EUSSET)
36. Guo Y, Liang C (2016) Blockchain application and outlook in the banking industry. Financ Innov 2(24):1–12. https://doi.org/10.1186/s40854-016-0034-9
37. Haber S, Stornetta S (1991) How to time stamp a digital document. In: Lecture notes in computer science, pp 437–455 (Advances in Cryptology—CRYPT0' 90)
38. Harris JD, Waggoner B (2019) Decentralized and collaborative AI on blockchain. In: 2019 ieee international conference on blockchain (Blockchain). IEEE, pp 368–375
39. Hasselgren A, Kralevska K, Gligoroski D, Pedersen SA, Faxvaag A (2020) Blockchain in healthcare and health sciences—A scoping review
40. Henry R, Herzberg A, Kate A (2018) Blockchain access privacy: challenges and directions. IEEE Secur Priv 16(4):38–45
41. Hoy MB (2017) An introduction to the blockchain and its implications for libraries and medicine. Med Ref Serv Q 36(3):273–279
42. Hughes L, Yogesh YK, Misra SK, Rana NP, Raghavan V, Akella V (2019) Blockchain research, practice and policy: applications, benefits, limitations, emerging research themes and research agenda. Int J Inf Manag 49:114–129

43. Hull R, Batra VS, Chen YM, Deutsch A, Heath FFT, Vianu, V. (2016). Towards as hared ledger business collaboration language based on data-aware processes. In: Sheng QZ, Stroulia E, Tata S, Bhiri S (eds) ICSOC2016. LNCS, pp.18–36. https://doi.org/10.1007/978-3-319-46295-0_2 9936
44. Hussien HM, Yasin SM, Udzir SNI, Zaidan AA, Zaidan BB (2019) A systematic review for enabling of develop a blockchain technology in healthcare application: taxonomy, substantially analysis, motivations, challenges, recommendations and future direction. J Med Syst 43(10):1–35
45. Hölbl M, Kompara M, Kamišalić A, Nemec Zlatolas L (2018) A systematic review of the use of blockchain in healthcare. Symmetry 10(10):470
46. Karafiloski E, Mishev A (2017) Blockchain solutions for big data challenges: A literature review. In: 17th International Conference on Smart Technologies (The IEEE EUROCON), pp 763–768
47. Kattwinkel O, Rademacher M (2020) Technical fundamentals of blockchain systems. BRSU communication report. https://pub.h-brs.de/frontdoor/deliver/index/docId/5005/file/BRSU_Communication_Report_1_Technical_Fundamentals_of_Blockchain_Systems.pdf. Accessed 19 July 2021
48. Katuwal GJ, Pandey S, Hennessey M, Lamichhane B (2018) Applications of blockchain in healthcare: current landscape & challenges. ArXiv, abs/1812.02776; Guo Y, Liang C (2016) Blockchain application and outlook in the banking industry. Financ Innov 2(24):1–12. https://doi.org/10.1186/s40854-016-0034-9
49. Kuo TT, Kim HE, Ohno-Machado L (2017) Blockchain distributed ledger technologies for biomedical and health care applications. J Am Med Inform Assoc 24(6):1211–1220
50. Lamport L (1998) The part-time parliament. ACM Tran Comput Syst 16(2):133–169. https://dl.acm.org/citation.cfm?doid=279227.279229
51. Li X, Jiang P, Chen T, Luo X, Wen Q (2017) A survey on the security of blockchain systems. Future Gener Comput Syst
52. Liang X, Shetty S, Tosh D, Kamhoua C, Kwiat K, Njilla L (2017) Provchain: a blockchain-based data provenance architecture in cloud environment with enhanced privacy and availability. In: 2017 17th IEEE/ACM international symposium on cluster, cloud and grid computing (CCGRID). IEEE, pp 468–477
53. Liebenau J, Elaluf-Calderwood SM (2016) Blockchain innovation beyond bitcoin and banking. https://ssrn.com/abstract=2749890Accessed 16 July 2021
54. Lindman J, Rossi M, Tuunainen VK (2017) Opportunities and risks of blockchain technologies in payments: A research agenda. In: 50th Hawaii International Conference on System Sciences, pp 1533–1542
55. Liu S (2021) Blockchain—Statistics and facts. Statista. https://www.statista.com/topics/5122/blockchain/
56. McClean K, Cross M, Reed S (2021) Risks to Healthcare organizations and staff who manage obese (Bariatric) patients and use of obesity data to mitigate risks: a literature review. J Multidiscip Healthc 14:577
57. Monrat AA, Schelén O, Andersson K (2019) A survey of blockchain from the perspectives of applications, challenges, and opportunities. IEEE Access 7:117134–117151
58. Moore T, Christin N (2013) Beware the middle man: Empirical analysis of bitcoin exchange risk. In: International conference on financial cryptography and data security, pp 25–33
59. Morabito V (2017) Business innovation through blockchain. Springer International Publishing, Cham
60. Mukhopadhyay U, Skjellum A, Hambolu O, Oakley J, Yu L, Brooks R (2016) A brief survey of cryptocurrency systems. The 14th annual conference on privacy, security and trust. IEEE, pp 745–752
61. Nakamoto S (2008) Bitcoin: a peer-to-peer electronic cash system. https://bitcoin.org/bitcoin.pdf
62. Narayanan A, Bonneau J, Felten E, Miller A, Goldfede S (2016) Bitcoin and cryptocurrency technologies: a comprehensive introduction. Princeton University Press

63. Nawari NO, Ravindran S (2019) Blockchain technology and BIM process: review and potential applications. J Inf Technol Constr 24:209–238
64. Niranjanamurthy M, Nithya BN, Jagannatha S (2019) Analysis of Blockchain technology: pros, cons and SWOT. Clust Comput 22(6):14743–14757
65. PAXOS (2021) Instinet and credit suisse conduct same-day settlement of traded stocks in historic first with PAXOS settlement service. https://www.paxos.com/instinet-and-credit-suisse-conduct-same-day-settlement-of-traded-stocks-in-historic-first-with-paxos-settlement-service/
66. Paul A, Haque Latif A, Amin Adnan F, Rahman RM (2019) Focused domain contextual AI chatbot framework for resource poor languages. J Inf Telecommun 3(2):248–269
67. Peng L, Feng W, Yan Z, Li Y, Zhou X, Shimizu S (2020) Privacy preservation in permissionless blockchain: a survey. Digit Commun Netw
68. Peter Y (2017) Regulatory issues in blockchain technology. J Financ Regul Compliance 25(2):196–208
69. Pilkington M (2015) Blockchain technology: Principles and applications. https://doi.org/10.4337/9781784717766.00019
70. Reportlinker (2021) Blockchain AI market research report by technology, by component, by deployment mode, by application, by vertical, by region—Global forecast to 2026—Cumulative IMPACT of COVID-19. https://www.reportlinker.com/p06081991/Blockchain-AI-Market-Research-Report-by-Technology-by-Component-by-Deployment-Mode-by-Application-by-Vertical-by-Region-Global-Forecast-to-Cumulative-Impact-of-COVID-19.html?utm_source=GNW
71. Rodrigues UR (2018) Law and the blockchain. Iowa L Rev 104:679
72. Rueda R, Šaljić E, Tomić D (2020) The institutional landscape of blockchain governance. A taxonomy for incorporation at the nation state. TEM J 9(1):181–187
73. Salah K, Rehman MHU, Nizamuddin N, Al-Fuqaha A (2019) Blockchain for AI: review and open research challenges. IEEE Access 7:10127–10149
74. Sankar LS, Sindhu M, Sethumadhavan M (2017) Survey of consensus protocols on blockchain applications. In: International conference on advanced computing and communication systems, pp 1–5
75. Sarmah SS (2018) Understanding blockchain technology. Comput Sci Eng 8(2):23–29
76. Sarpatwar K, Sitaramagiridharganesh Ganapavarapu V, Shanmugam K, Rahman A, Vaculin R (2019) Blockchain enabled AI marketplace: the price you pay for trust. In: Proceedings of the IEEE/CVF conference on computer vision and pattern recognition workshops
77. Seebacher S, Maleshkova M (2018) A model-driven approach for the description of blockchain business networks. In: Proceedings of the 51st Hawaii international conference on system sciences
78. Seebacher S, Schüritz R (2017) Blockchain technology as an enabler of service systems: a structured literature review. In: The 8th international conference on exploring service science, pp 12–23
79. Sharma PK, Park JH (2018) Blockchain based hybrid network architecture for the smart city. Futur Gener Comput Syst 86:650–655
80. Sharples M, Domingue J (2016) The blockchain and kudos: a distributed system for educational record, reputation and reward. In: European conference on technology enhanced learning. Springer, Cham, pp 490–496
81. Skiba DJ (2017) The potential of blockchain in education and health care. Nurs Educ Perspect 38(4):220–221
82. Subramanian H, Cousins K, Bouyad LB, Sheth A, Conway D (2020) Blockchain regulations and decentralized applications: panel report from AMCIS 2018. Commun Assoc Inf Syst 47(1):9
83. Sun J, Yan J, Zhang KZK (2016) Blockchain-based sharing services: what blockchain technology can contribute to smart cities. Financ Innov 2(26):1–9. https://doi.org/10.1186/s40854-016-0040-y

84. Syed TA, Alzahrani A, Jan S, Siddiqui MS, Nadeem A, Alghamdi T (2019) A comparative analysis of blockchain architecture and its applications: Problems and recommendations. IEEE access 7:176838–176869
85. Szostek D (2019) Blockchain and the law. Nomos Verlag
86. Tama BA, Kweka BJ, Park Y, Rhee KH (2017) A critical review of blockchain and its current applications. In: International Conference on Electrical Engineering and Computer Science, pp 109–113
87. Tapscott A, Tapscott D (2017) How Blockchain is changing finance. Harv Bus Rev 1(9):2–5
88. The Economist (2015) The great chain of being sure about thing. https://www.economist.com/news/briefing/21677228-technology-behindbitcoin-lets-peoplewho-do-not-know-or-trust-each-other-build-dependableAccessed 19 July 2021
89. Truby J (2018) Decarbonizing bitcoin: law and policy choices for reducing the energy consumption of blockchain technologies and digital currencies. Energy Res Soc Sci 44:399–410
90. Van Alstyne M (2014) Why bitcoin has value. Commun ACM 57(5):30–32
91. Viriyasitavat W, Hoonsopon D (2019) Blockchain characteristics and consensus in modern business processes. J Ind Inf Integr 13:32–39
92. Werbach K (2018) Trust, but verify: why the blockchain needs the law. Berkeley Tech LJ 33:487
93. Xu JJ (2016) Are blockchains immune to all malicious attacks? Financ Innov 2(25):1–9. https://doi.org/10.1186/s40854-016-0046-5
94. Yaga D, Mell P, Roby N, Scarfone K (2018) Blockchain technology overview. arXiv:1906. 11078. https://nvlpubs.nist.gov/nistpubs/ir/2018/NIST.IR.8202.pdf. Accessed 19 July 2021
95. Yeoh P (2017) Regulatory issues in blockchain technology. J Financ Regul Compl 25(2):196–208
96. Yermack D (2017) Corporate governance and blockchains. J Rev Financ 21(1):7–31
97. Yin HHS, Langenheldt K, Harlev M, Mukkamala RR, Vatrapu R (2019) Regulating cryptocurrencies: a supervised machine learning approach to de-anonymizing the bitcoin blockchain. J Manag Inf Syst 36(1):37–73. https://doi.org/10.1080/07421222.2018.1550550
98. Yuan Y, Wang FY (2018) Blockchain and cryptocurrencies: model, techniques, and applications. IEEE Trans Syst Cybern: Syst 48(9):1421–1428
99. Yuan Y, Wang FY (2016) Towards blockchain-based intelligent transportation systems. In: 2016 IEEE 19th international conference on intelligent transportation systems (ITSC). IEEE, pp 2663–2668
100. Zhao JL, Fan S, Jiaqi Yan YJ (2016) Overview of business innovations and research opportunities in blockchain and introduction to the special issue. Financ Innov 2(28):1–7. https://doi.org/10.1186/s40854-016-0049-2
101. Zheng Z, Xie S, Dai H, Chen X, Wang H (2017) An overview of blockchain technology: architecture, consensus, and future trends. In: 2017 IEEE international congress on big data (BigData congress), IEEE, pp 557–564
102. Zyskind G, Nathan O (2015) Decentralizing privacy: using blockchain to protect personal data. In: 2015 IEEE security and privacy workshops. IEEE, pp. 180–184

Fundamentals of Blockchain and New Generations of Distributed Ledger Technologies. Circular Economy Case Uses in Spain

Romero-Frías Esteban, Benítez-Martínez Francisco Luis, Nuñez-Cacho-Utrilla Pedro Víctor, and Molina-Moreno Valentín

Abstract The evolution of Distributed Ledger Technologies since the first blockchain constitutes an opportunity for digital transformation in many social and economic contexts. This transformation will become even more intense in the post–covid-19 economic scenario. One of the most significant changes in the transformation consumers have undergone in their perception of the environment and of the sustainability of our societies. The circular economy is an agent of change that helps minimize humankind's impact on the planet. And this change must be aligned with the sustainable development goals of the 2030 Agenda. Since its inception, blockchain technology has represented an opportunity for transparency, immutability, and persistence in the processes in which it intervenes, allowing it to be incorporated into easily traceable, disintermediated networks. By implication, this suggests we can move towards a verifiable, reliable, circular economy. In this chapter we cover the fundamentals of blockchain technology including its history, the elements that define distributed ledger technologies, and how they have evolved

R.-F. Esteban (✉)
Department of Accounting and Finance, Faculty of Economics and Business, University of Granada, Campus Cartuja, s/n, 18071 Granada, Spain
e-mail: erf@ugr.es

R.-F. Esteban · B.-M. Francisco Luis
MediaLab, University of Granada, Avda. Madrid, s/n, Espacio V Centenario, 18071 Granada, Spain
e-mail: flbenitez@fidesol.org

B.-M. Francisco Luis
FIDESOL, BIC Building, Avda. Innovación, s/n. Health Technology Park (PTS), 18100 Armilla, Granada, Spain

N.-C.-U. Pedro Víctor
Polytechnic School of Linares, University of Jaen, Ronda Sur s/n, Linares, 23700 Jaen, Spain
e-mail: pnunez@ujaen.es

M.-M. Valentín
Department of Management-1, Faculty of Economics and Business, University of Granada, Campus Cartuja, s/n, 18071 Granada, Spain
e-mail: vmolina2@go.ugr.es

S. S. Muthu (ed.), *Blockchain Technologies for Sustainability*,
Environmental Footprints and Eco-design of Products and Processes,
https://doi.org/10.1007/978-981-16-6301-7_2

into their most recent forms—e.g., Hashgraph, Direct Acyclic Graphs, Holochain or Neural Distributed Ledgers—which represent scalable, efficient, innovative solutions aligned with sustainability goals and facilitating the efficient, sustainable management of circular economy projects. Finally, drawing on striking use cases in Spain, we describe the use of blockchain technology in circular economy projects, and its operating status, as examples of ideas for the future development of models that may reach the market.

Keywords Blockchain · Distributed ledger technologies · Spain · Use cases · Circular economy · Neural distributed ledgers · Sustainability · Applications

1 Introduction

It is more than a decade since Satoshi [22] defined the primal blockchain: bitcoin. The promise of a currency that questioned centralized financial systems through disintermediation, eliminating the intermediate authorities that centralized trust, was a challenge. Moreover, the principles of transparency and cooperation, on which the traceability of the transactions performed between the nodes and their immutability and temporary sealing are based, are the essential elements of this technology and can contribute most to the development of a full circular economy (CE).

The aim of this chapter is to provide as complete and concise an overview as possible of the basics of blockchain technology and DLTs so as to provide the lay reader with an effective entry point, which we address in the context of the CE. Blockchain and DLT are economic and technological concepts that set new paradigms which have entered the economic and social mainstream in parallel over the last 10 years. In particular, we focus on the fundamentals of blockchain technology: historical blockchain issues; definitions; key concepts; "actors" in a DLT; constituent elements of blockchain; governance conditions for future applications; and, finally, a review of new generation blockchains like Hashgraph, Direct Acyclic Graphs (DAG), Holochain, Chia Network, DFinity, and Neural Distributed Ledger (NDL).

We conclude by reviewing some applications of blockchain to the CE in Spain in order to demonstrate how this novel technology can be implemented to solve real, practical economic issues in real scenarios.

2 History and Evolution of Blockchain Technology

When Satoshi [22]—whoever he may be—revealed his intentions in the article in which he unveiled the first blockchain, bitcoin, he was not really just defining new data recording technology. He proposed a paradigm shift in the way that we determine how to originate and exchange the monetary value of money. With bitcoin, the first cryptocurrency was born, and with it, a whole new sector: fintech.

The bitcoin genesis blog from January 2009 contains the following message: "The Times 03/JAN/2009 Chancellor on brink of second bailout in weeks for banks".

Quite a declaration of intent, right in the middle of the Great Recession that began in 2008. The blockchain emerged within an entire ecosystem with more political than financial leanings, although these views are two sides of the same "coin" [12]. More than a decade later, society at large has yet to fully understand blockchain technology. Perhaps, its initial adherence to crypto-financial systems and a bad press—since it was first used in the most obscure corners of the "Dark Web"—have jeopardized its social and economic implications.

Let us begin by asking how much trust is worth and how much it costs. These questions are inherent to the economic or social transactions that any of us have to face on a daily basis. Trust is the "glue" of society. It is the axis on which social institutions are built and, as such, it is the lintel that supports the social architecture of our civilizing process. Consequently, the major revolution that accompanies blockchain is the way in which it (re)configures trust management since it establishes a central criterion: "disintermediation", that is, the need, or rather the non-need to have "third parties" as intermediaries who confer trust, one of the most precious, essential, social values—particularly in our digital times.

Blockchain technology stands on several pillars. One is the generation of a distributed, disintermediated system without nodes or central entities that decide who has access and who does not. Another is the development of a cryptographic system that supports this. Cryptography has been with us from Antiquity right up to the more recent work of Alan Turing during World War II, and with the appearance of blockchain, it has now acquired special relevance in cybersecurity systems. On this, the creation of bitcoin was based.

And, once again, as so often before in technological advances, the germ of the technology is not in Satoshi Nakamoto's now-famous white paper but in the work of the Belgian cryptographer Jacques Quisquater, who in May 1999 together with Henry Massias and Xavier Serret Ávila, published an essay entitled "Design of a Secure Timestamping Service with Minimal Trust Requirements", within the TimeSec project. Their essay described just that: a timestamping system with minimal trust requirements, supported by a distributed system [18]. Quisquarter is also the creator of the GQ cryptographic scheme, which is well known in the cybersecurity systems sector.

Moreover, the technology has another co-creator: Ralph Merkle, one of the fathers of public key algorithms and whose research was transcendental in the birth of the blockchain. Without his "hashes tree" to build block transactions, the technology would not have hatched [20].

Nowadays, "DLTs" constitute a vast, complex ecosystem that has multiple definitions and, according to literature reviews, is quite inconsistent. The same lack of terminological standardization affects this and all blockchain technology ([25]: 11).

The concept of DLT predates the existence of bitcoin and the blockchain technology powered by cryptocurrency. Lamport et al.'s [16] theoretical Problem of the Byzantine Generals described a hostile environment in which computer systems had to handle conflicting information reaching them from different sources. Subsequent research led to the development of the first algorithm that tolerated "Byzantine failures" in high availability systems, with little increase in latency [9].

In fact, before Nakamoto's white paper [22], two studies had identified elements of what would later become the blockchain: the notion of a chain of cryptographic blocks linked by blocks that ensured the sealing of digital data in distributed systems, using cryptographic hashing functions; and Merkle trees [6, 13].

As in the history of computer science, before the emergence of a new technology or paradigm shift, such as the Internet, earlier work paved the way towards the development of new, disruptive models. The current abundance of DLTs, with different configurations and types, often makes it difficult to establish a clear taxonomy of how they operate and how they are constituted.

The concept has been established as a general term to designate multi-party systems functioning in an environment with no central operator or authority, even though the parties involved may be unreliable or malicious, and may exist in harsh environments. Blockchain technology is considered a specific subset of the broader DLT ecosystem, and it uses a specific data structure consisting of a chain of data blocks linked with cryptographic hashing functions. The concept of DLTs was first described in 1982, and that of blockchain first appeared in 1991. However, they were actually deployed prior to their massive incorporation into society.

Consequently, we need to clarify the fact that a hostile environment in a DLT is characterized by the presence of malicious actors in the system or network, who undermine it by subverting its intended use. The prototypical adversary in a DLT system is an entity that attempts to exploit consensus rules to transfer assets without authorization, censor the transactions of others, or disrupt or destroy the network. Opponents can operate both inside (on-chain) and outside (off-chain) the system. Hence, governance schemes to establish the management framework are crucial to the management of any type of platform [8].

The learning curve demanded by this technology remains long, both in the academic field and in its translation into effective market solutions. As previous research has suggested, one potential application of blockchain is in the field of the CE, particularly in relation to regulatory implications and their impact on its deployment [14]. Its potential and possibilities are extensive, as suggested by ([19, 32]: 362–370) and, most recently ([3]: 429–440). However, the starting point is starkly clear: many blockchains will coexist, and many others will be defined in the coming years, representing a challenge in terms of new opportunities and of barriers against and complexities in solving current and future problems in the CE.

3 A Definition of Blockchain

A blockchain is a distributed database, which is shared and agreed on in a peer-to-peer network. It consists of a linked sequence of blocks, containing a timestamp (for each block) and transactions secured by a public cryptographic key and verified by the entire network community. Once an item is added to the blockchain, it cannot be altered and becomes an immutable record of past activity.

The absence of standardization and the difficulty in limiting the minimum parts of a distributed registry ecosystem have meant that they can be classified in many different ways. From an academic point of view, researchers have proposed several classifications of blockchain technology. For instance [23], put forward a two-dimensional approach: by typology of the existence of the authority that predefines it, and in terms of how the nodes that cooperate in it are encouraged to participate. Platt [24] shared this binary approach since it focuses on the blockchain data dissemination model and the functionality of its on-chain system ("stateful" versus "stateless"). Lemieux [17] analyzed blockchain from an archival point of view and proposed the following types of record keeping system: the "mirror type", the "digital record" and the "tokenized model". Lemieux also examined each type in relation to a formal framework of theoretical evaluation of defined file types. Xu et al. [30] based their classification on a study of the approximation of the layers available to each type of blockchain. Their aim was to evaluate the impact of design decisions in each blockchain on the architecture of the software that defines them. Their proposed taxonomy examined aspects of software architecture relating to the performance and quality of the systems studied. Finally, we would like to point out two studies published in 2018, which place the accent on the concepts studied earlier, and conduct a structured review of these. Tasca and Tessone [27] attempted to determine a comprehensive, detailed ontology of previous research. Ballandies et al. [5] conducted a technological review and constructed a taxonomy of 29 systems chosen from the more than 1000 existing at the time. To date, their work constitutes one of the best summaries of DLT typologies and their characterization with a view to future attempts at standardization.

Following Rauchs et al.'s classification [25], to be recognized as such, a DLT must at least exhibit the following properties:

(a) A shared registry system allowing multiple parties access to the data, in order to create, maintain and update them in a single collective data set: the "ledger".
(b) A multi-party consensus system, allowing the network and each party to reach an agreement on the data recorded in the "ledger".
(c) If this is not allowed, it does not depend on a single party or previous parallel agreements, and there is an absence of ex-ante trust relationships between the parties; if it is allowed, it occurs through multiple record producers that have been approved and bound by some form of contract or other types of agreement, between the participating nodes.
(d) It must have an independent validation system that allows each participating node to independently verify the status of the stored transactions and the collective integrity of the system.
(e) Intrusion evidence is one of the most robust properties of a DLT. It enables each node to detect non-consensual changes in the network that have been applied trivially, in violation of the governance system, to validate data or network transactions.

(f) Resistance to tampering makes it impossible for a single node to unilaterally change a record or the transaction history of blocks.

Therefore, a DLT's main task is to produce a set of authoritative records that are validated and executed through a multi-party consensus process that involves the participation of multiple separate nodes. And all which takes place in a disintermediated way, in the clear absence of any central authority. Network users create and transmit unconfirmed transactions (i.e., proposals to make a new entry in the general ledger), which the record producers group together for aggregation. Once confirmed, the instructions contained in the transactions are automatically executed by all network auditors, to be incorporated into the DLT.

4 Key Concepts

The concept of "ledger" is central to the technology and implies an understanding of the transformative capacity that blockchains entail [31]. However, it should be noted that in the academic field two different ideas tend to overlap. On the one hand, "ledger" is defined as the set of data that most nodes have in common; on the other, it is the set of data held by a node in an individual network. We believe that this dual vision actually operates as one which, in our judgment, would be the authorized set of collective records of a significant proportion of network participants at any one time so that the records are unlikely to be deleted or modified (i.e., "definitive"). And one very important factor that must be taken into account is the fact that it is an abstract construct that enables us to understand the collective power of the distributed network that fosters and maintains a DLT. It is not an object that exists independently, nor is it unique or stored independently. This level of abstraction is what determines the format of this technology.

In order to facilitate novel readers' understanding of this technology, here is a selection of the most commonly used terms relating to a DLT:

- Native assets: the primary digital assets, if any. They must be specified in the protocol that is normally used to regulate the production of records, pay transaction fees on the network, carry out "monetary policy" (if any) or align incentives to maintain platform integrity.
- Consensus algorithms: the set of rules and processes used by the network to reach agreement and validate the final records.
- 51% attack: the type of attack in which a subset of DLT participants, with a majority of the governance system votes, produces registrations faster than the other participants. This can lead to a situation in which, when the data produced in this way is revealed to the rest of the network, those recorded by the "honest" nodes may be replaced due to the conflict that arises from any modification of the rules that originate them. This type of attack is the oldest (and classic) against a DLT that operates Proof of Work (PoW)-type consensus mechanisms.

- Multi-party consensus: the ability of the system to allow independent participants to agree on a shared set of records without requiring a central authority.
- Tamper evidence: participants' ability to easily detect arbitrary changes in confirmed records.
- Fork: the event in a DLT whereby it splits into two (or more) networks. A fork can occur when two or more record producers more or less simultaneously publish a set of valid records as part of an attack, or when a protocol change is forced on the system.
- Journal: the set of records that contains/owns a node. Diaries are provisional and highly heterogeneous, depending on the governance typology of the platform on which the transactions are created. They may or may not contain all the same records.
- Ledger: the platform's set of authorized records that, at any point in time, is collectively owned by a significant proportion of the network participants. This record is unalterable, and cannot be amended or deleted.
- Log: an unordered set of validated transactions, which contains a node, that has not yet been incorporated as a formal data record, subject to the consensus rules of the system; i.e., the set of unconfirmed transactions of a node.
- Node: a participant in the DLT in communication with other "peers" through a shared communication channel for the production of the records.
- Off-chain: the set of interactions, actions, and processes that occur outside the formal limits of the DLT.
- On-chain: the set of interactions, actions, and processes that occur within the system (i.e., at system level) and are reflected in the data layer.
- Persistence: the ability of data to remain available after program execution and to survive catastrophic loss of an arbitrary number of nodes, whatever their origin.
- Protocol: the set of rules defined by the software that determines how the system works.
- Record: the set of recorded data of transactions that have been subject to the platform's consensus rules. Before being incorporated, certain Log records are called "candidate records", as an intermediate step, and their main characteristic is that they have not been propagated by the network in which they are inserted.
- Permissioned Network: a network in which a record can only be created by a specific set of participants.
- Permissionless Network: a network in which a candidate record can be created by any participant.
- Endogenous Reference: data that can only be created and transferred through the system and only has meaning within the system. Execution is carried out automatically by the system itself.
- Exogenous Reference: data that refer to some condition of the physical world and must be incorporated from the outside. This generally requires a gateway to connect to the external system and enforce decisions outside of the DLT.
- Hybrid Reference: data that share endogenous and exogenous characteristics. Their execution partly depends on the gateways that affect them and bring them to completion.

- Candidate Record: a registry that has not yet been propagated to the network and that therefore is not subject to the internal consensus process of the nodes, as indicated above.
- Tamper resistance: the ability to make it difficult for a single party to unilaterally change past records (i.e., the history of ledger transactions).
- Transaction Executed Programmatically: a script that, when activated by a particular message, is automatically executed by the system. When code is capable of operating as all parts intend, the deterministic nature of execution reduces the level of trust required for individual participants to interact with each other. They are commonly referred to as "smart contracts" due to the ability of scripts to substitute code for certain fiduciary relationships, such as custody and escrow of information. However, they are neither autonomous nor contracts in a legal sense; rather, they should be understood as the technological means to automatically implement a contract or agreement once pre-established conditions have been manifested.
- Transaction: any proposed change in the "ledger", regardless of the type of data that is recorded; it does not necessarily have an intrinsic economic connotation. In fact, it is an event registered in the "ledger".
- Independent Validation: the ability of the system to allow each participant to independently verify the status of their transactions and the integrity of the system itself.
- Validation: this is the set of processes necessary to ensure that the actors independently reach the same conclusion regarding the status of the ledger. It includes verifying the validity of unconfirmed transactions, verifying registration proposals, and auditing the status of the system.

5 The "Actors" in a DLT

Another term that can be confusing is the use of "actor" in a DLT. "Actor" typology differs according to the role each participant plays in the processes. It can refer to a person or an entity that interacts with the system. According to the literature, actors can be divided into many categories, according to the role they play in the system. It is worth noting that an entity can adopt several roles at the same time, depending on its components, within each layer of the system.

- Developers are in charge of writing and reviewing the code that builds the DLT and its interconnecting systems. They may be responsible for maintaining the code base of the central protocol that supports the system. They can also be in charge of designing applications (dApps) that can be executed in the DLT or of creating the infrastructures that enable the protocols that work between them.
- Administrators control access to the system's central code and decide who they can add. They are normally in charge of controlling and executing the governance of the system. In a DLT, their function can vary greatly: it can be public and permissionless, or private and permissioned.

- Gateways are basically entities that help the DLT fulfill its tasks in developing the processes transacted by its components. They are a "bridge" between the DLT and the outside world and can serve multiple purposes.
  - Gatekeepers guarantee participant access to the network.
  - Oracles are in charge of transmitting the external data to the network.
- Custodians are entities in charge of keeping the assets in custody.
- Exchanges are responsible for facilitating the purchase/sale of digital assets. They are indispensable in any DLT that starts with an Initial Coin Offering (ICO), to build a cryptocurrency.
- Issuers are in charge of issuing or exchanging tokens that represent the assets registered in the system.
- Participants: the network consists of a series of interconnected participants (often referred to as nodes) that criss-cross information and messages to build the DLT ledger. They can take on many roles. They can be auditors, checking the transactions and records that are chained; independent nodes in the system; or they can produce records based on the log for their final validation in the ledger. And, in turn, they can be end-users, as custodians of the token storage systems (Wallet System).

6 The Constituent Elements of a Blockchain

The potential of this technology is emerging thanks to the promise of its being secure and tamper-proof with digital records. This allows us to foresee its impact as enabling technology in the face of the challenges posed by the 4th Industrial Revolution.

Specifically, blockchain comprises seven noteworthy characteristics [15], Ben [4]:

- Decentralized system. In blockchain there is no need for transactions to be validated through a central trusted agency; this eliminates inefficiencies and any dependency on a central structure. The existence of a third party is not necessary to make the system reliable.
- Consensus algorithms. These are used to maintain the consistency of the data and the robustness of the distributed networks used by the platform.
- Persistence. Transactions can be validated quickly because so-called "honest miners" will not accept invalid transactions. It is practically impossible to delete or alter any transaction that has been incorporated as a record in the blockchain. Blocks containing invalid transactions can be discovered immediately.
- Algorithmic trust. Trusted transactions between network peers that have no other basis on which to trust each other are derived from almost incorruptible encryption, which validates all current and historical transactions on each blockchain.
- Auditability. Any transaction must refer to some previous unverified transactions. Once the current transaction is recorded on the blockchain, the status of those

referred transactions changes from unverified to verified. Thus, transactions can be easily verified and tracked across the entire network.
- Immutability. In minutes or even seconds, all transactions made are verified, erased, and stored in a block that is linked to the previous block, thus creating a chain.
- Public. Anyone can view it at any time because it resides on the network, not within a single institution charged with auditing transactions and keeping records.

7 Governance Conditions for Future Applications

When designing a DLT there are several issues to analyze, depending on the path you want to choose. Every choice creates demands on the system that can no longer be waived.

Choosing the system governance has potential implications regarding its management decision-making processes and operating rules (Alston 2019). In addition, it must be recalled that this decision will affect the system's sustainability and its strength, as well as the actors' perception of legitimacy, derived from its degree of transparency and how to gain access to the network from abroad. This whole set of derivatives will determine which model is chosen and, when this occurs, any change or possible solution will depend on that initial choice.

A priori, we must take account of the possible governance model configurations available [7] because this choice determines everything, as we outlined above. Different types exist, depending on the configuration protocol:

- Anarchic: with models of collaboration and cooperative decision-making, on a totally voluntary basis, without any type of central authority (or validating nodes). This model tends to establish contentious relationships that are permanently discussing protocols and rules with a high degree of failure or network division. Bitcoin fits this model perfectly.
- Hierarchical: participants have the right to propose changes, but there is a central committee or entity (made up of validating nodes) that manages the network rules and protocols. This governance is typical of permissioned private blockchains.
- Federated Model: only a group of agents can vote on alterations to the platform's protocols and rules, but those who participate in this process are in a horizontal hierarchy, although they may not carry the same weight in terms of the platform's votes. This model is becoming increasingly important because it allows for a distributed system with a higher rate of efficiency, transaction speed, and scalability than the anarchic model.
- Democratic Model: all participants have the right to vote, conditioned by their weight and capacity on the platform. Predefined rules establish how and under what circumstances decisions are made.
- Dictatorial Model: a governance model that is at the limit of what we can consider a DLT. It requires a central entity among the participants (which they consent to) that determine the protocol. It is very common in blockchains dedicated to mining

all kinds of cryptocurrencies. Originating in bitcoin and Ethereum, it is a way to guarantee that there is no fork in the system.

This choice will also determine the type of network access. That is, whether it is open, closed, or semi-open. This will imply the typology of participants in the system, as well as the type of consensus mechanism from which the trust system between the nodes will derive. These decisions have implications on the actors' participation rights, platform maintenance costs (and the type of reward, for which the nodes participate in it), and the degree of resistance when the blocks are sealed.

We could go on to highlight other potential implications but these are at the core of all existing DLT platforms. Designing a DLT platform implies an initial decision-making process that determines its entire useful life cycle and what it can be dedicated to. Given that this is still an experimental technology, to achieve progress in its future development, certain initial paradigms—characteristic of the two most widespread platforms: bitcoin and Ethereum—must be left behind in order to move towards others that allow us to overcome the problems we outlined earlier.

Another issue permanently associated with a DLT is managing the security of the records stored in it [29]. The question of how records are validated, verified, and sealed remains the great unknown to be answered, so as to ensure resilient platforms, secure against cyberattacks, as reflected in the most recent study published by [21].

In previous sections, we have tried to identify the elements we consider essential in identifying a DLT. This is especially important since this technology is a disruptive element in business environments, with a conceptual model that transcends outdated approaches to solving emerging problems in our society [19]. Proposals are needed for conceptual tools that help legislators and developers lay the foundations for next generation DLTs. Understanding the interactions between the processes of the different layers of the technology and how these affect the design of future potentially commercial solutions is one of the most pressing needs that the scientific literature should address in the coming years, to propose new solutions to the major global challenges facing society.

8 New Generations of Blockchain

After the birth of bitcoin, the first blockchain, the Ethereum network was launched: blockchain 2.0, bringing with its smart contracts. Much of the technological solutions they provide have been discussed above. But today's DLTs challenge the architecture on which the two best-known versions were built.

Evolution and new developments are constant. Depending on how we conduct transactions and store their data, unique ecosystems are created that bring progress on two of the major obstacles to the use of DLTs in the marketplace: scalability and the increase in transactions per second.

8.1 Hashgraph

This is a DLT that is based on the creation of consensus, in particular timestamping consensus, to ensure that transactions on the network match each node on the platform. This type of consensus algorithm highlights the robustness and superiority of DLT networks.

Unlike a traditional blockchain, users do not have to submit PoW. This means that the nodes do not have to validate the transactions that take place on the network. This eliminates the need for two elements: firstly, the DLT does not need to perform many calculations to achieve a successful transaction, which implies a high rate of transactions per second. Secondly, Hashgraph only requires that the network nodes reach consensus by the "Gossip about Gossip" method (through which all nodes can validate and process information and, at the same time, have specific knowledge of all details of the operation with a virtual voting technique), which is similar to what we have seen in the most recent generation of decentralized oracles, such as that used in Astraea (Adler et al. 2018). The sum of both processes achieves a highly impartial consensus system. This means that there is very little time between the start and end of a transaction, which has a direct implication on the efficiency of the network itself. Stress tests published by the Hashgraph team (pending an external audit) claim the network can reach 10000 transactions per second (TPS). Additionally, it uses an extremely low transaction fee by comparison with bitcoin or Ethereum and has been prepared so that it can make micropayments.

8.2 Dag

This DLT is based on directed acyclic graphs that use a totally different data structure to those in use since the first blockchain. It relies on a consensus mechanism, whose algorithm is designed so that all nodes cooperate with each other and all have the same rights. Like Hashgraph, this DLT guarantees the impartiality of the network between the nodes.

Its promoters decided to turn system governance effort into a democratic system among equals, without main nodes or node categories. Tangle by IOTA and ByteBall is the first DLTs to use this architecture, which is still at an early stage of development and will potentially launch experimental designs that move away from the traditional structure of a typical blockchain that we have seen in this chapter. One of its great innovations, which has significant implications in its governance system and the algorithm of its consensus protocol, is that a node can issue and validate a transaction in the same unit of time.

8.3 Holochain

This is another DLT that goes outside the concept of the blockchain architecture that we have studied. In fact, it has been designed to lead a new generation of blockchain. According to its white paper, it is intended to revolutionize the Internet and its management methods, surpassing the framework on which the server-client relationship is built and moving towards a fully distributed system.

Its aim is to create a distributed network on which the future generation of the Internet will run and guarantee access, democratization, and internal freedom. Its promoters define it as an amalgam of blockchain, BitTorrent, and GitHub. The idea is clear: distribute the data flow of the "network of networks" across its nodes. In addition, according to its governance system, its nodes will have the freedom to operate autonomously, and the data they share will be distributed via different locations around the world. This design frees the network from the risk of system load congestion, which automatically makes it highly scalable. In theory, once operational, flows of millions of transactions per second could be achieved. Moreover, if a developer wanted to deploy a dApp on this system, they would only need to confirm the single chain of the entire DLT, i.e., the DLT itself. If all this can be put into practice, it could change the worldwide management of digital information.

8.4 Chia Network

This blockchain was presented in February 2021. In its white paper, it is distinguished from the best-known blockchains in two ways. Firstly, it is based on a system that takes advantage of the computer's disk storage instead of incurring high energy costs due to the use of computer graphic resources to mine the tokens. This makes PoW unnecessary. Instead, it uses a consensus mechanism called Proof of Space and Time (PoST) [1] to take advantage of the data management capabilities of hard drives. This is where the blockchain's native token—the "Chia"—is mined. It develops an intelligent transaction platform based on the UTXO model. The UTXO approach is much simpler, easier to implement, has lower overhead on full nodes, and results in much more reliable, secure, and succinct smart transactions. The Solidity model (bitcoin) is much more dangerous, expensive, and unreliable, but it does have greater expressive power. Chia's developer is Bram Cohen, the inventor of BitTorrent, the most widespread P2P system on the Internet and the clear inspiration behind this DLT.

The original plan was to create a new generation of cryptocurrencies, with a native language to program intelligent transactions. This language is called Chialisp and is designed to be easily auditable and to easily integrate the PoST consensus mechanism. Given that it was born as a new option on the financial market until it is widespread and demonstrates its capabilities as a sustainable DLT and its transaction rates per second, it will be impossible to analyze its potential use.

8.5 DFinity

This kind of DLT is developed by a Swiss foundation that has designed the Internet Computer Protocol (ICP) with a self-governing model. It combines special node machines run "en masse" by independent parties from independent data centers around the world. It works like a normal blockchain and the code used is tamper-proof.

In fact, they use a new concept of smart contract called "canisters". These canisters can directly serve end-users with web content and interact with blockchain services, without holding tokens. This approach focuses more on a smart contracted platform than a tokenizable model. So their dApps can be developed simply by uploading canisters into cyberspace. The canisters are pre-charged with "cycles" that fuel their computation, so end-users do not pay, unlike on the Ethereum platform.

Dfinity uses the ICP tokens converted into cycles; through computing routines, they "burn" cycles in a circular process. This allows it to ignore an ICO, creating a model in which developers can raise funds by selling the internal governance tokens created in the canisters.

Finally, the ICP utility tokens enable participation in an algorithmic, open governance system, called the Network Nervous System, which securely manages and updates the network (in addition to being the source of cycles that power computation). To participate in this, an ICP token holder locks them inside "voting neurons". Voting power and rewards derive from key attributes, including the total number of tokens locked, the dissolve delay (the delay before locked tokens are released after dissolving has begun), and neuron age. Increasing commitment, by increasing the dissolve period, increases power and rewards. Neurons are non-transferrable.

8.6 Neural Distributed Ledger (NDL)

Beyond the DFinity proposal, we find one more DLT. The NDL concept is a collaborative, multidimensional blockchain network. Its internal design is analogous to the way that groups of neurons are structured in the human brain. Hence the term: "neural distributed ledger". It operates by adding all "memories" (transactions) stored in the clusters that make up the platform. It works like a "ledger ledger", as defined by [28].

The idea of establishing a neural model to develop a blockchain like "chains of personal thought" was defined by [26] as a bridge towards the integration of Artificial Intelligence and the Internet of Things, to develop new applications and new types of architecture as universal transaction frameworks. The trade name for this DLT is RETIS, and it is a first step towards a deeper understanding of the visionary concept described by Swan.

It is specifically designed to meet the requirements of business and government environments that need to consume and process massive data in their management processes. It is a private, permissioned network blockchain that hosts mixed public and private information in all its nodes. It acts as a 3D network that accumulates nodes in different blockchains within the platform while building blocks in the blockchains that it contains and manages in a federated way. This ensures the network's vertical and horizontal scalability, which forms a 3D mesh to deal with the massive management of data and transactions that the platform must support per unit of time. It does not need to use consensus mechanisms, which makes it a highly sustainable network, and it develops a totally different typology of Byzantine fault tolerance. To do this, it uses specific algorithms for network governance in a totally on-chain way.

9 The Concept of Token: "Tokenization"

In recent years, hundreds of cryptocurrencies have come onto the market. However only a very few of them obtain a trading base with fiat money in the market. Most have been implemented on the Ethereum platform and manage the mining of their cryptocurrencies through smart contracts that generate a token. Each one can present different properties, according to the rules defined in their respective governance system white papers. The standard token is the ERC-20, which is also the most popular in Ethereum. Chen et al. [10] reviewed the ICOs based on this token—more than 80% of the total—which serves to show the importance of this standard on the platform.

The differences between cryptocurrency and token must be clearly understood. These terms are often confused due to the influence of fintech solutions in the current blockchain market. Cryptocurrencies are the form of digital money created by blockchain solutions whereas the token represents an asset or utility with a specific tangible or intangible value in the community that created it. These are usually transferable goods that can range from loyalty points, game bonuses, or future rights for a service that can be exchanged when an agreed result occurs.

Clearly, it follows that each has a different structure and purpose. Principally, cryptocurrencies originate within their own blockchain, while tokens always operate "above" their blockchain. Furthermore, cryptocurrencies always have a monetary value purpose, while tokens can represent any type of asset that is useful in society; this does not have to be a monetary value but could, for example, be the casting of a vote in an electoral process.

A token can be defined as a digital asset that operates "on top" of a cryptocurrency or a blockchain and is often executed as a programmable asset thanks to a smart contract, to be used within a project or a dApp.

When we consider that cryptographic tokens represent the right to something, we are defining the tokenization of a digital asset. Tokenization is a way of converting the rights of something or someone, into a digital artifact, which acquires the digital format of a token. With cryptographic tokens, the benefits of tokenization reside mainly in their greater versatility, greater liquidity, improved programming capacity, and the fact that they constitute immutable proof of ownership [11]. However, there is still a great lack of tokenization standards and, above all, most countries lack a legal infrastructure and legal framework that regulates and defines the concept of tokenization.

As a medium of exchange, tokens can act as currency themselves. In this sense, they can also be used as the local currency of a dApp. In general, cryptographic tokens are used beyond mere economic exchange by taking advantage of their most prominent feature: the fact that they are programmable. This became possible thanks to Ethereum. The first cryptographic token on a blockchain, bitcoin, does not offer this.

Since they are potentially programmable, they are used to activate certain functions in dApp smart contracts. Additionally, tokens can be tied to off-chain assets. They can serve as a means of fund raising, reserve, or investment, as well as being a way to build an ecosystem or community, which can use them to exchange rights or obligations that have been predefined by their users.

The value of a token depends mainly on the supply, demand, and trust that the participating community place in it, which in turn are based on the credibility, trust, and service it offers the community.

In the absence of a legal or academic framework that establishes a clear taxonomy of tokens, they can be grouped into three well differentiated classes:

(a) Payment tokens (for cryptocurrencies, essentially)
(b) Security tokens
(c) Utility tokens.

The main distinguishing feature is the investment purpose of the security tokens (for crowdfunding or the search for investors, for example) as opposed to the added value in the operation of a product or service that is typical of utility tokens. Payment tokens, having no other pre-established function, are born in the ICOs of the cryptocurrencies to which their value is associated. Security tokens are identified as "assets", such as a debt or a capital right of the issuer, whereas utility tokens are generally associated with a project or a dApp with a tangible benefit and provide digital access to a specific application or service thanks to a blockchain-based infrastructure.

The basic functionality of a token contract comprises accounting for who owns the tokens, transferring token ownership by agreeing to changes in the wallet in the respective token contract, and managing events to register transfer of ownership in the ledger records. Secure transfer is a mechanism whereby tokens are withdrawn from an address after approval (by nodes) instead of being transferred to an address where they might be lost if it were not ready to receive them. There are other specific functionalities relating to the creation (minting) and destruction (burning) of tokens, and their distribution and trade (e.g., through an ICO). Furthermore, token contracts often implement general functionalities including authentication functions and node roles, control (e.g., pause, lock), and the provision of information to users (e.g., display functions).

10 Uses and Applications. The Spanish Case

We now present examples of the use of blockchain technology in the circular economy in Spain.

10.1 Climate Trade

Climate Trade is an organization[1] that helps companies achieve carbon neutrality by offering a series of emission compensation services based on a blockchain traceability structure. It has developed a marketplace in which any company can decide on a portfolio of international cooperation projects to offset its carbon footprint. The projects operate in both voluntary and mandatory markets. They are classified by territory and by 2030 Agenda SDGs so that users can filter them easily. They have three basic types of projects: Carbon Offsetting Projects (neutral carbon fingerprint projects); Nature-based Solutions (sustainable projects); and Guarantees of Origin (advanced sustainable projects).

Climate Trade has its own API, which allows it to offer the compensation service in a simple, transparent way thanks to a dashboard that supervises and controls all transactions conducted on the platform. The procedure is simple and the system requires that users follow three steps:

1. During the purchase process, the user calculates the exact carbon footprint generated by the product or service they are going to use.
2. They obtain the cost of the footprint and, from the portfolio, choose which project they want to use to compensate for it.
3. The user receives an automatic email with a nominative certificate with all the project information.

Climate Trade is currently administering their carbon footprint compensation system for companies and individuals such as Iberia, Telefonica, Acciona, Banco Santander, Cabify, IAG International Airlines Group, Suez, Danone, and Meliá Hotels.

They hope to run a CE project involving large-scale operators so that individuals and companies can offset their carbon footprint by participating in traceable sustainability projects in the custody of their blockchain solution.

10.2 Telos-Based CTIC Logistics Traceability Application

The CTIC (Information Technology Center Foundation) technology center has an ongoing project to ensure traceability in the food chain. Its objective is to give

[1] https://climatetrade.com/.

consumers certain guarantees about the legitimate origin, production, and distribution of food in accordance with current legislation.

The Telos blockchain platform is a third generation blockchain and builds scalable, distributed applications with feeless transactions and on-chain governance. It is managed by the Telos Foundation. The platform has 21 "active" validators who are voted in by TLOS holders. They take primary responsibility for maintaining the network, while a cohort of "standby" producers is paid and regularly tested to act as backup.

The TLOS holders can vote any of the standby validators into the top 21 at any time via elections that are held continuously, approximately every 2.5 min. Equally, producers can be voted out of the top 21 at any time too. The network also has standby validators that are automatically rotated into the top 21 at intervals to give their operations a chance to produce blocks and prove their readiness.

The Telos network has a dStor decentralized solution with an InterPlanetary File System (IPFS) in its stack. It facilitates the creation of Distributed Autonomous Organizations (DAOs) and, with the Telos Etherium Virtual Machine, can create solidity smart contracts to be deployed in the platform.

The CTIC has designed two apps that interact with data recorded in Telos. The first is a consultation application for the final consumer of the product who wants to know more about its origin and characteristics. The second is a data registration application for all companies involved in the chain of production, processing, distribution, and sale of the product.

This is to ensure the integrity of the supply chain and protect the goods against counterfeiting and overproduction, in addition to providing the end consumer with information of interest.

10.3 Food Track, Food Supply Chain Traceability

Developed by the Spanish start-up Nutrasign,[2] this platform allows users to create a unique, secure, immutable digital record of each product, offering traceability from origin to table. It identifies a production batch and its path in seconds. The designers have tried to respond to the four main challenges in traceability:

(a) Greater transparency in the supply chain.
(b) Compliance with import and export regulations.
(c) Mitigating consequences of outbreaks of disease or contamination.
(d) Improving brand confidence and exposure.

Their solution is complemented by a Food Journey that certifies the history of the food. It validates the origin and processes and also ensures the authenticity of products throughout the supply chain. It is a solution that enhances consumers' increasing concerns over the origin of their food. Consumers are interested in brands

[2] https://www.nutrasign.io/.

that provide sustainability, care for the environment, and comply with corporate social responsibility.

10.4 CircularChain, a Blockchain Platform for Packaging Recycling

This solution has been built by Ecoembes[3] and Minsait[4] (Indra Group) and constitutes the main CE blockchain platform in Spain. This CE platform's network of distributed registers will help public administrations, local entities, operators, recyclers, and other organizations to safely share and control all system data and accelerate all transactions relating to the waste selection process while promoting process transparency, awareness, and cost reduction.

It will enable the agile implementation of smart auditing systems based on the records generated by all those involved in the chain. This will make it clear that corporations and governments comply with their environmental commitments to waste management and generate environmental control systems.

CircularChain constitutes a single source of truthful information and an unalterable record, making all operations and transactions conducted by participants in the ecosystem more transparent and more easily auditable. Furthermore, as there is no central system, the platform has no global infrastructure development or maintenance costs; these are shared among the participants.

CircularChain will facilitate the recording of unchangeable information on traceability (withdrawal identifier, plant, material type, or possible non-conformity, among others) and documentary evidence that guarantees the existence of a document at a given time and facilitates its inspection for later modifications. In addition, it enables the establishment of intelligent agreements, for example, on the delivery of materials or packaging, which facilitate the automation of payments or penalties for non-compliance.

It also records hundreds of daily transactions associated with the removal of material from over 90 sorting plants and transfers from over 70 recyclers right across Spain. In November 2020, a pilot project on the blockchain platform recorded over 3700 withdrawals and some 14500 transactions.

This technology allows public administrations to facilitate access to audited, unalterable data on CE processes and their integration into a network that encourages public-private collaboration.

[3] https://www.ecoembes.com/en/home.

[4] https://www.minsait.com/en.

11 Conclusions

In this chapter, we have outlined the main characteristics of a distributed ledger technology. However, there remains a lack of understanding as to what does or does not constitute a DLT. Moreover, an academic and economic debate also exists about what comprises a blockchain or a DLT. Interestingly though this is, in the present chapter we have avoided analyzing this topic, instead of arguing that blockchain represents both the first and second generations of DLTs. A blockchain is just one type of DLT. In Sect. 7, we defined the most compelling proposal for new, 4th generation DLTs, which entails a change of mindset if they are to be adopted in CE projects.

Subsequently, we have described the most exciting, stimulating projects currently deployed in Spain—many of which use Ethereum. Clearly, we would encourage the use of current, widely used blockchain platforms to develop dApps and dApp-based solutions. However, if we want Agenda 2030 to make headway, we need to design and use new tools and frameworks that could fully match CE projects.

We need to use a Green DLT in any future solution. We understand that the DLTs described in this chapter—such as NDL and Dfinity—have undoubted potential to overcome the challenges implied by SDGs.

With the present study, we hope to draw attention to how future blockchains aligned with Agenda 2030 should be used to fulfill the philosophy of the CE and sustainability.

References

1. Abusalah H, Alwen J, Cohen B, Khilko D, Pietrzak K, Reyzin L (2017) Beyond Hellman's time-memory trade-offs with applications to proofs of space. In: International conference on the theory and application of cryptology and information security. Cham, Springer, pp 357–379
2. Alston E (2019) Constitutions and Blockchains: Competitive Governance of fundamental rule sets. SSRN Electronic Journal. https://doi.org/10.2139/ssrn.3358434
3. Attaran M, Gunasekaran A (2019) Blockchain-enabled technology: the emerging technology set to reshape and decentralise many industries. Int J Appl Decis Sci 12(4):424–444. https://doi.org/10.1504/IJADS.2019.102642
4. Ayed AB, Belhajji MA (2018) The blockchain technology: applications and threats. Int J Hyperconnectivity Internet Things IGI Glob 1(4):1–11. https://doi.org/10.5555/3213107.3213108
5. Ballandies MC, Dapp MM, Pournaras E (2021) Decrypting distributed ledger design—taxonomy, classification and blockchain community evaluation. Cluster Comput 1–22. https://arxiv.org/abs/1811.03419. Accessed 7 May 2021
6. Bayer D, Haber S, Stornetta WS (1993) Improving the efficiency and reliability of digital time-stamping. In: Capocelli RM, De Santis A, Vaccaro U (eds) Sequences II: methods in communication, security, and computer science. Proceedings of the sequences workshop. Springer, New York, pp 329–334. https://doi.org/10.1007/978-1-4613-9323-8_24
7. Benedict G (2019) Challenges of DLT-enabled scalable governance and the role of standards. J ICT Standard River Publ 7(3):195–208. https://doi.org/10.13052/jicts2245-800X.731

8. Brown AE, Grant GG (2005) Framing the frameworks: a review of IT governance research. Commun Assoc Inf Syst (Carleton University) 15(38). https://doi.org/10.17705/1cais.01538
9. Castro M, Liskov B (2002) Practical Byzantine fault tolerance and proactive recovery. ACM Trans Comput Syst (TOCS) 20(4):398–461. https://doi.org/10.1145/571637.571640
10. Chen W, Zhang T, Chen Z et al (2020) Traveling the token world: a graph analysis of Ethereum ERC20 token ecosystem. In: Proceedings of the web conference 2020, pp 1411–1421. https://doi.org/10.1145/3366423.3380215
11. Di Angelo M, Salzer G (2020) Tokens, types, and standards: identification and utilization in Ethereum. In: 2020 IEEE international conference on decentralized applications and infrastructures (DAPPS). Oxford, United Kingdom, pp 1–10. https://doi.org/10.1109/DAPPS49028.2020.00001
12. Golumbia D (2016) The politics of bitcoin: software as right-wing extremism. University of Minnesota Press
13. Haber S, Stornetta WS (1991) How to time-stamp a digital document. J Cryptol (Berlin, Heidelberg, Springer) 3:99–111. https://doi.org/10.1007/BF00196791
14. Hacker P, Thomale C (2018) Crypto-securities regulation: ICOs, token sales and cryptocurrencies under EU financial law. Eur Company Financ Law Rev 15(4):645–696
15. Kan L, Wei Y, Muhammad AH et al (2018) A multiple blockchains architecture on inter-blockchain communication. In: ieee international conference on software quality, reliability and security companion (QRS-C). Lisbon, Portugal, pp 139–145. https://doi.org/10.1109/QRS-C.2018.00037
16. Lamport L, Shostak R, Pease M (1982) The byzantine generals problem. ACM Trans Progam Lang Syst 4(3):382–401. 10.1.1.12.1697
17. Lemieux VL (2017) A typology of blockchain recordkeeping solutions and some reflections on their implications for the future of archival preservation. In: 2017 IEEE international conference on big data (Big Data), pp 2271–2278. https://doi.org/10.1109/BigData.2017.8258180
18. Massias H, Avila XS, Quisquater JJ (1999) Design of a secure timestamping service with minimal trust requirement. In: The 20th symposium on information theory in the Benelux, pp 79–86
19. Maull R, Godsiff P, Mulligan C et al (2017) Distributed ledger technology: applications and implications. Strateg Change. Brief Entrep Financ 26(5):481–489. https://doi.org/10.1002/jsc.2148
20. Merkle RC (1988) A digital signature based on a conventional encryption function. In: Advances in cryptology—CRYPTO '87. Lecture notes in computer science. Berlin, Heidelberg, Springer, pp 369–378, 293. https://doi.org/10.1007/3-540-48184-2_32
21. Moubarak J, Chamoun M, Filiol E (2020) On distributed ledgers security and illegal uses. Futur Gener Comput Syst 113:183–195. https://doi.org/10.1016/j.future.2020.06.044
22. Nakamoto S (2008) Bitcoin: a peer-to-peer electronic cash system. https://nakamotoinstitute.org/bitcoin/
23. Okada H, Yamasaki S, Bracamonte V (2017) Proposed classification of blockchains based on authority and incentive dimensions. In: 2017 19th international conference on advanced communication technology (ICACT), pp 593–597. https://doi.org/10.23919/ICACT.2017.7890159
24. Platt C (2017) Thoughts on the taxonomy of blockchains & distributed ledger technologies. Medium. https://medium.com/@colin_/thoughts-on-the-taxonomy-of-blockchains-distributed-ledger-technologies-ecad1c819e28. Accessed 5 Aug 2020
25. Rauchs M, Glidden A, Gordon B et al (2018) Distributed ledger technology systems: a conceptual framework. SSRN. https://ssrn.com/abstract=3230013
26. Swan M (2015) Blockchain thinking: the brain as a decentralized autonomous corporation. IEEE Technol Soc Mag 34(4):41–52. https://doi.org/10.1109/MTS.2015.2494358
27. Tasca P, Tessone CJ (2018) Taxonomy of blockchain technologies. Princ Identif Classif 31. https://doi.org/10.2139/ssrn.2977811
28. Velasco C, Colomo-Palacios R, Cano R (2020) Neural distributed ledger. IEEE Softw 37(5):43–48. https://doi.org/10.1109/MS.2020.2993370

29. Workie H, Jain K (2017) Distributed ledger technology: implications of blockchain for the securities industry. J Secur Oper Custody 9(4):347–355
30. Xu X, Weber I, Staples M et al (2017) A taxonomy of blockchain-based systems for architecture design. In: 2017 IEEE international conference on software architecture (ICSA). Gothenburg, pp 243–252. https://doi.org/10.1109/ICSA.2017.33
31. Zhang K, Jacobsen HA (2018) Towards dependable, scalable, and pervasive distributed ledgers with blockchains. In: IEEE 38th international conference on distributed computing systems (ICDCS), Vienna, Austria, pp 1337–1346. https://doi.org/10.1109/ICDCS.2018.00134
32. Zheng Z, Xie S, Dai HN et al (2018) Blockchain challenges and opportunities: a survey. Int J Web Grid Serv 14(4):352–375. https://doi.org/10.1504/IJWGS.2018.095647

Security Magnification in Supply Chain Management Using Blockchain Technology

Bharat Bhushan, Anushka, Abhishek Kumar, and Lucky Katiyar

Abstract In supply chain management the necessity for data transparency is very essential as it is key to create trust between retailers and customers. But data managed by centralized controllers face several vulnerabilities and security threats like data breaches, data confidentiality, and many more. Blockchain is a digital and distributed ledger has acquired great popularity in recent years, due to its security, immutability, and transparency in data. It solves many challenges like keeping the data secure by using cryptographic algorithms. It is a decentralized ledger for recording, managing, storing, and transmitting data in a peer-to-peer network. This paper aims to provide a brief survey on the magnification of security in supply chain operations using blockchain, further indicating the challenges encountered during the integration. The work presents a descriptive study of past literature on blockchain for intensifying security in supply chain operations by examining the features provided by blockchain technology. Further, the paper provides an insight into how blockchain is transforming the business by providing safe and automated solutions. Additionally, this paper highlights the motivation behind using blockchain technology in supply chain management. Further, the work investigates how leveraging blockchain can help in overcoming vulnerabilities and avoiding fraudulent activities in the traditional supply chain. Finally, the paper highlights the uses of the blockchain-based business and enumerates the related future research directions.

Keywords Blockchain · Supply chain management · Smart contract · Decentralized and distributed ledger · Logistics · Security · Data integrity

1 Introduction

Supply Chain Management (SCM), is the process of managing the flow of services and goods, from transforming raw materials into a final product reaching customers [1]. It comprises various steps of a product life cycle and usually requires the

B. Bhushan (✉) · Anushka · A. Kumar · L. Katiyar
Department of Computer Science and Engineering, School of Engineering and Technology, Sharda University, Uttar Pradesh, India

S. S. Muthu (ed.), *Blockchain Technologies for Sustainability*,
Environmental Footprints and Eco-design of Products and Processes,
https://doi.org/10.1007/978-981-16-6301-7_3

cooperation of several businesses and stakeholders, which makes the supply chain extraordinarily complex and vulnerable too [2]. Additionally, the complexity of the SCM increases drastically in international trading scenarios. Many intermediaries are involved at each stage and each activity is conducted according to the documents shared by the third party regulating the whole trade which also results in mistrust. The complexity of supply chains is aggravated by factors like extra delays, disruption, information distortion, lack of transparency, increased costs, and other uncertainties. But now due to technological transformations happening around the world, there are significant security driven solutions being implemented to optimize the SCM [3].

Introduced by Santoshi Nakamoto, the hype around bitcoin and blockchain technology has been acquiring tremendous attention around SCM. Blockchain is a shared and decentralized ledger for recording, managing, storing, and transmitting data in a Peer-to-Peer (P2P) network [4]. It makes use of cryptographic algorithms forming a structure like a chain of interconnected blocks containing the data [5]. Blockchain provides a sustainable and secure architecture for the operation of the supply chain. IBM, Walmart, and many other companies are investing heavily in this field to improve supply chain processes by harnessing the power of blockchain technology in various industries [6]. The interest of the research communities is also helping blockchain technology gaining momentum in the last few years which is evident by a series of literature reviews [7]. Blockchain technology is inherently distributed, decentralized, and tamperproof making it a potential solution to address the issues in traditional SCM [8].

The main focus of this paper is to discuss the disruption of traditional SCM leading to digitization of supply chain operations which is rather more advanced and secure. The traditional Logistics and Supply Chain Management (LSCM) has many limitations like misusing the data shared between the parties, illegally sourcing the products, etc. Adoption of solutions powered by blockchain ensures addressing all these limitations and seamless flow of activities at each stage and counterfeiting fake transactions and reducing paperwork. The technical emphasis of this paper is on leveraging blockchain to enhance security, traceability, responsiveness, and give real-time information of the products. Studies on the application of blockchain in almost every industry are it the coffee manufacturing industry or the automotive industry, show the importance, and capabilities of blockchain technology to establish a sustainable, simplified yet secure supply chain management.

In summary, the major contributions of our work are listed as follows:

- This work presents a thorough study of blockchain technology highlighting the various types, characteristics, and key features.
- This work presents a comprehensive survey of supply chain operations in traditional SCM and explores the existing vulnerabilities therein. It also outlines the evolution of traditional supply chain and the effect of technological transformations.
- This work presents the motivation to use blockchain in supply chain operations and highlights its intervention in improving supply chain operations.

- Finally, this work presents an organized and practical survey of harnessing blockchain in supply chain processes. It further provides the description and applications of this technology in various supply chain industries.

The rest of the paper is organized as follows. Section 2 presents relevant literature review. Section 3 gives the overview of blockchain technology, types of blockchain; private, permissioned, consortium, and hybrid along with the differentiation of each, characteristics, and the key components. Section 4 describes SCM, the motivation behind using blockchain technology in SCM. It further explains the digitization of SCM, advantages of using blockchain technology, along the use cases in various sectors. Finally, the paper concludes in Sect. 5 highlighting several future research directions in the field.

2 Literature Review

To make sure a secure system is established in blockchain, Hafid et al. [9] proposed a work that gives solutions to keep the failure probability smaller than a predefined threshold for a sharding protocol in blockchain. They used three probability bounds: Chvatal, Chebyshev, and Hoeffding illustrating the effectiveness of the model they proposed. They also conducted a comparative and numerical analysis of the bounds proposed. Belotti et al. [10] presented how the application of blockchain is beyond bitcoin, surveying numerous literature of the past few years. Indicated requirements, evolution from private to public blockchains, and listing the differences between proposed and consensus mechanisms. Tsoulias et al. [11] presented a decentralized application model which stores the data in Neo4j graph database assisting protocol operations and enhancing security. They implemented a consensus mechanism similar to Casper and tested its effectiveness. They used Proof-of-Work (POW) and Proof-of-State (POS) protocols to examine how incentive and consensus criteria differ for participants. Further, they also conducted a series of experiments that tested the efficiency of the implemented solutions and methodologies to prevent the most common 51% attack, which is an attack from dynamic validator sets and catastrophic crashes as well.

To ensure the seamless flow of operations in supply chain Asyrofi et al. [12] proposed a system that is based on cloud that manages supply chains using selective marketing and blockchain. They also improvised Jugo architecture to develop Selective Market (SELAT) as a connector between cloud providers and the users. It also improves data security by tracking the changes in the supply chain by using blockchain. Rouhani et al. [13] presented a structured review focusing on smart contracts and how it has widened the horizons of blockchain's application beyond bitcoin and other cryptocurrencies. They laid down their research study in three major categories: decentralized applications which are based on smart contracts, security tools and methods, and approaches to enhance the performance of the smart contracts.

Hader et al. [14] presented an introduction of blockchain in supply chain management in the retail sector. They also offered a comprehensive study on how companies can improve their performance and build trust with their customers using blockchain technology. Reyes et al. [15] presented a work discussing the impact of blockchain, Internet of things (IoT) in the operation of SCM. They identified the benefits and implications of leveraging these technologies in a multi-organizational supply chain setting. They also discussed how technology can accelerate businesses making the process transparent and cost-efficient. Hassija et al. [16] proposed at work discussing application of blockchain and other technologies to achieve secure trade. They have also discussed the solutions to vulnerabilities in the existing architecture of the supply chain. Fu et at. [17] described an intelligent operating mechanism and system structure to be applied in large production enterprise supply chains. They also constructed a data success and storage mechanism further providing a model structure for developing a blockchain-based supply chain. Musigmann et al. [18] presented a work that fills the gap by implementing a bibliometric and co-citation analysis in blockchain technology and LSCM. They classified the past literature into five different categories: testing and conceptualizing blockchain applications for the operation of LSCM and the role of blockchain in digitization of supply chains. Wu et al. [19] presented a work that focuses on deployment of blockchain in future networks and vertical industries. They discussed how blockchain is being implemented in several sectors such as supply chain, finance, energy due to its ability to create a transparent and tamperproof nature. They also discussed the potential of blockchain as a solution to achieve security in laying networks. Tran et al. [20] presented a detailed study of blockchain outlining the challenges related to privacy and security further classifying the areas of application to enhance security. They enlisted several areas of application such as data management, e-voting systems, smart agriculture, cryptocurrency, etc. They also proposed a framework called Privacy Preserving Blockchain Systems (PPSAF) designed specifically to resolve the issues in agriculture industry. Finally, outlining the scope of future research. Table 1 presents a systematic overview of the related literature.

3 Overview of Blockchain

Blockchain was invented in 2008 to serve as a public distributed ledger for bitcoin. Blockchain is a specific type of database, unlike the traditional relational database. It stores the data in the blocks which is encrypted by a digital signature called hash which makes it more reliable and secure, preventing the data from tampering in a network [21]. The first block is called the genesis block and each block is linked together hence forming a chained structure [22]. Each of these blocks consists of the hash value of the previous block, hash of its own, the transaction data, and the nonce value [23]. The hash value is an alphanumeric unique digital signature that is used to identify the block. These hashes are generated by using SHA-256 algorithms. Number only used once, also abbreviated as a nonce, is the number the

Table 1 Overview of related literature

Reference	Year	Contribution
Hafid et al. [9]	2019	Sharding protocol in blockchain to keep the failure probability smaller
Belotti et al. [10]	2019	Indicated requirements, evolution from private to public blockchains, and listing the differences between proposed and consensus mechanisms
Tsoulias et al. [11]	2020	Designed decentralized application model which stores the data in Neo4j graph database
Asyrofi et al. [12]	2020	Created a cloud-based system that manages supply chains using selective marketing and blockchain
Rouhani et al. [13]	2019	Studied decentralized applications which are based on smart contracts, security tools, and methods and approaches to improve the performance of the smart contracts
Hader et al. [14]	2020	Introduced blockchain in supply chain management in the retail sector and offered a comprehensive study on how companies can improve their performance
Reyes et al. [15]	2020	Discussed impact of blockchain, Internet of things in the operation of SCM and identified the benefits along with implications of leveraging these technologies
Hassija et al. [16]	2020	Described application of blockchain to achieve secure trade
Fu et at. [17]	2019	Designed an intelligent operating mechanism and system structure for large production enterprise supply chains
Musigmann et al. [18]	2020	Implemented a bibliometric and co-citation analysis in blockchain technology and LSCM
Wu et al. [19]	2021	Deployment of blockchain in future networks and vertical industries
Tran et al. [20]	2021	Studied decentralized applications which are based on smart contracts, security tools and methods, and approaches to improve the performance of the smart contracts

blockchain miner needs to explore to solve the blockchain through mathematical calculations. The data stored in these blocks is immutable and irreversible [24]. Blockchain consists of the nodes which are typically managed by the P2P network. These nodes strictly adhere to the protocol to communicate and validate the new blocks which are added to the blockchain. As one block is altered in the chain, it breaks the cryptographic links hence disrupting the complete blockchain [25]. Decentralized blockchain has data that is immutable which means that the data is irreversible. The irreversible nature of blockchain is an advantage and hence making it fully transparent to the people in the network, which means people can easily track any information. The data in the blockchain can only be appended and cannot be changed or deleted. Blockchain has several benefits to offer; easy tracking, decentralization of data, low administrative costs help eliminate the vulnerabilities in the existing traditional systems [26].

3.1 *Types of Blockchain*

There are various types of blockchain networks based on the visibility of the nodes in the network, and how the users connect to the network. We have public, private, consortium, and hybrid blockchain and each type have its architecture, transparency, efficiency, and immutability.

3.1.1 Public Blockchain

Public blockchain as its name suggests is a permission-less or non-restrictive distributed ledger system that can be accessed by anyone on the internet, become part of the network, do the mining, view the records, verify the transactions, and do the POW for the new blocks. It is more secure because such a large network is hard to hack [27]. Also, one person is not dependent on the other for authentication, so this is more trustable. Although it has few drawbacks like transaction per second is high, this raises issues and high energy consumption. Anyone in the world can read, send the transaction, and participate in the consensus process. These are secured by crypto economics. Crypto economics is the combination of cryptography and economic incentives which provide decentralized and secured systems. Example: Bitcoin, Ethereum, Bitcoin Cash, Litecoin, Monero, IOTA.

3.1.2 Private Blockchain

Private blockchain is a permissioned or restrictive ledger operative only in a closed network and governed by a single entity. It can be used within the organization or enterprises where only a few members can be part of the network, view the record, and do the mining [28]. Users need to get access rights to the network to validate blocks and send transactions. The write permissions are restricted while read permissions can be public. As there are only a few nodes on the platform, the efficiency is always high in a private blockchain. Example: Multichain, Quorum.

3.1.3 Consortium Blockchain

It is unlike the public blockchain, is controlled by multiple organizations or enterprises [29]. In consortium blockchain there are fewer members and all of them are known. The members can read or write transactions but they cannot add a block. There is an exclusion of 51% attack in this type of network. For industrial applications, it is considered to be the most suitable. The transaction cost is reduced and replaced by the legacy systems, simplifies the process of handling documents, and also helps to get rid of semi-manual compliance mechanisms. Example: Hyperledger, Corda.

3.1.4 Hybrid Blockchain

It is the amalgamation of both private and public. It gives us freedom like that in public and permission access like that in private. The architecture of hybrid blockchain can be customized as per the needs and requirements [30]. Once a user gets permission to access the hybrid blockchain platform, he can completely take part in the activities like sending transactions, reading, or writing too. The identity of the users is kept secret for privacy reasons. It is immune to 51% attack and hackers cannot access the network. Additionally, it gives low transaction costs too. Example: Ripple, and XRP. Table 2 presents a categorized differentiation of each type of blockchain network.

Table 2 Comparison of types of blockchain networks

Distinguishing features	Public blockchain	Private blockchain	Consortium blockchain	Hybrid blockchain
Definition	It is open to all and anyone in the network can participate	It is controlled by administered owners and only members in the network can participate	It is controlled by a group of people, i.e.; multiple companies will have full control and administrative rights	It is a combination of both public and private which means some of the processes are kept private and have limited access while some are open to all
Authority and transparency	In public blockchain, all processes are completely transparent	Private blockchain is transparent to the users who are given the access	A group of people has access hence it is decentralized	Transparency in hybrid is dependent on how administrators set the rules
Transactional cost	It is costly	It is not so costly	It is not so costly	It is not so costly
Use cases	It can be used for any public project. It can also be used in creating cryptocurrencies	It can be implemented in an organizational project where the control is limited to its members while few processes are only accessed by its users	It can be implemented where no single organization has full authority	It is best suited for projects that have lack of trust

3.2 Characteristics of Blockchain

Some characteristics of blockchain technology that are helping developing businesses and organizations are listed below.

- Blockchain is designed in such a way that it is synchronized and distributed across the networks, due to which it provides an ideal system for multi-organizational industries networks like supply chains or financial consortia. It helps an organization to disclose its data and share it publicly through the feature of data transparency which helps label institutional deficiencies to achieve sustainability.
- The data is immutable and cannot be changed by any third party outside of the organization. The kind of transactions that are carried out is agreed upon between participants in advance. And these transactions are stored in blockchain also called smart contracts, which help in providing assurance that everyone is operating by the rules decided.
- Prior to the execution of a transaction by the participant, there needs to exist an agreement among all those relevant parties that the transaction which is to be done is valid. Example, if a person is registering for the sale of a product, that product must belong to a certain organization, or else it won't get approved. Such a process is known as consensus which helps keep the fraudulent transactions away from the database.
- Once the organization has accepted a transaction and registered it, the transaction can never be changed. They can consequently register another transaction concerning that particular product to change its status, but cannot hide the original or initial transaction that was made earlier. This provides the notion of the provenance of product, which also means that one can track and access the history of any asset and determine its present or past locations and what has happened throughout its life.
- Blockchain functions great in terms concerning cost-cutting benefits and technical considerations for an organization, which helps a business overcome the counterfeit and fake component products in their supply chains. It aids in sustainable supply chains for developing countries. Also, its transparency can profit less important members in supply chains [31].

3.3 Key Components of Blockchain

Blocks containing data are chained together to each other, forming a linked list that makes up the blockchain. Each block has a hash along with the timestamps. Figure 1 shows the structured view of blockchain including all the main components. All the key components and pillars are discussed as follows:

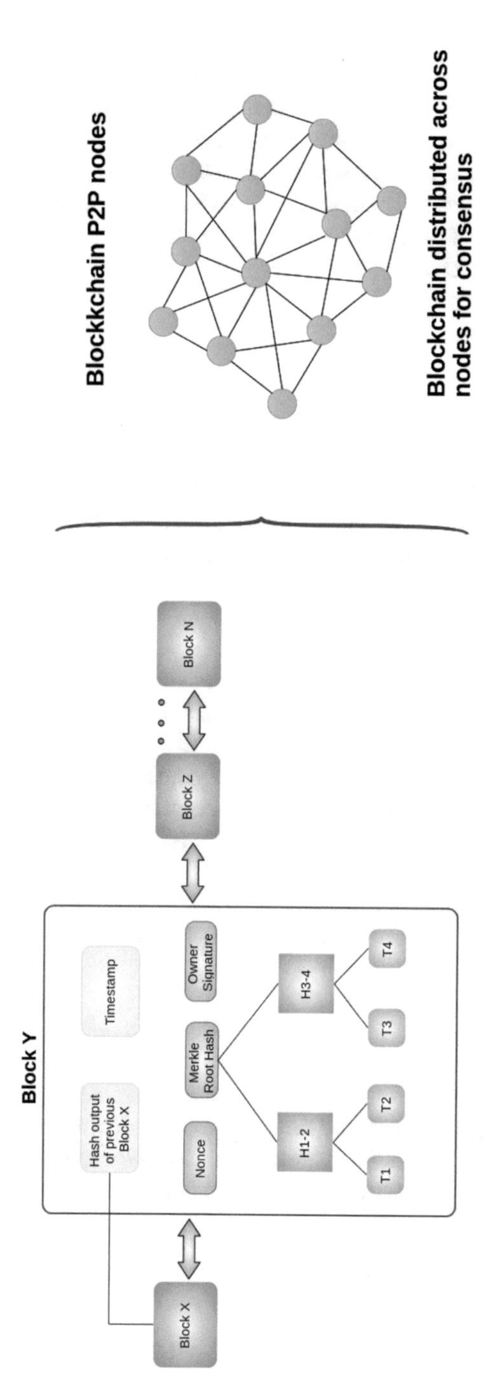

Fig. 1 Overview of blockchain components

3.3.1 Block

A block is a data structure consisting of transaction records in chronological order [32]. It contains a hash pointer that points to the previous block. A block is made up of a header that contains metadata and list of transactions. The block header has a size of 80 bytes and the size of the average transaction is approximately 250 bytes while an average block generally contains at least 500 transactions. The initial block is called the genesis block. And as soon as the creation of a block is done the data contained becomes verifiable.

3.3.2 Chain

One block is linked to the previous block, forming a sequence called a chain similar to a linked list. The local copy of the blockchain keeps altering as new blocks are joined. As soon as the node obtains a block from the network, it validates them and links to the blockchain [33]. The integrity of the chain can be determined from the very first block that is the genesis block to the last one.

3.3.3 Transaction

These are the smallest data structures of a blockchain. It is created by users to indicate the transfer of data from a user to a specific receiver [34]. Before a transaction is added to the blockchain it is authenticated through cryptographic keys and authorized by the users. In a blockchain, the decision of adding a transaction is done using consensus protocols, which means nodes have to agree if the transactions occurring in the network are genuine and valid.

3.3.4 Nodes

The blockchain is managed by a P2P network where the nodes collectively cohere by the protocols to validate and communicate new blocks. These network nodes are completely responsible for storing the data that enters and sharing information about transactions. Blocks of data are stored on the nodes which are usually computers or servers connected to each other. Node's store and save blocks of transactions. They spread and broadcast the transaction histories to other nodes to synchronize with the blockchain [35].

3.3.5 Miners

The miners create new blocks through a process known as mining. Mining is the process of validating every step in the transaction which is achieved by mathematical computations. These are the specific nodes performing the process of block verification before adding anything to the blockchain. Without a full node, miners cannot know what are valid transactions [36]. The role of miners is to find the hidden encryption code called POW. It is challenging but rewarding at the same time. Mining is useful for securing the network, validating and confirming transactions.

3.3.6 Merkle Trees

Each block in the blockchain aggregates the brief of all the transactions using the Merkle tree. It is also called a binary hash tree and it's a structure used for effectively verifying and summarizing the data. It uses the double-SHA256 cryptographic algorithm [37]. Merkle tree sets apart the validation of the data from the data itself and helps to maintain the functional integrity of the transactions. They allow compressing of the large data-keeping while only keeping the ones required to prove.

4 Supply Chain Management

SCM is the handling of the flow of products, goods, and services, including every process that transforms raw materials into a final product. Traditionally, SCM seeks to cater consumer's demands for a service or product along with the slightest inventory required for the retailer and producer [4]. Many supply chain models were proposed and designed earlier during the 80 s when the term SCM was coined, striving to satisfy the needs of numerous manufacturing networks. Some of the requirements include the reduction of costs from inventories and transactions, eliminating bottlenecks along the supply chain due to delays in supply delivery and payments. It also included the creation of chains resilient to fluctuations due to limited primary material and poor economic condition, the traceability of commodities sourced in a trustful and secure way, employing of local laborers and producers, minimizing the transportation needs, and distribution of high-grade quality products for the end consumer. The supply chain binds the end-to-end movement, including correlated and physical flow of products, raw materials, information, and money. The supply chain is involved in managing the sourcing, distribution, production, procurement, and logistics, thus, influencing the expense of a product, speed-to-market, required capital, and service perception in businesses [38].

4.1 Motivation Behind Applying Blockchain Technology in SCM

Blockchain technology has been an influential conversation point in the supply chain industry. Even today several supply chain leaders are bewildered and dubious about the future of launching blockchain technology in the real world. They may understand its scope, but implementing this technology in the real world seems out of reach. Due to this, supply chain executives may not apprehend that investing in blocks and technology is the only method to mitigate these shifts in standard operating procedures successfully. But the use of blockchain technology in the supply chain is assumed to be the end-all solution to transparency and visibility. The complexity of the modern supply chain is extensive but blockchain technology has the potential to fix problems within today's supply chain networks. Major organizations, like Maersk and IBM, are struggling on creating innovative platforms to allow for end-to-end transparency and visibility through this technology. The expectation of launching technology means saving billions through the elimination of electronic data interchange and paper-based systems, reducing inefficiencies, vulnerability, inconsistency, and other major drawbacks. Products may slip between modes, and a large global manufacturing present exacerbates the complexity. Which increases the risk of purchasing counterfeit products, the "gray market". Through blockchain technology, consumers can check the product they purchased, and vendors, merchants, and suppliers can take the same action. This increases efficiency, transparency, and reliability within the supply chain. Supply chain companies using blockchain in the supply chain can examine the chain of ownership and deliver an extra layer of security and guarantee, to other companies regarding the item in question's value and validity. These are essential parts in supply chains that have a direct connection to the health of the public, such as the food, agriculture, and medical industries.

The implementation of the supply chain depends on every faction of numerous activities which are termed as supply chain practices. These practices involve several elements that comprise every aspect of the supply chain processes; those include sharing of information, integration, outsourcing, customer relationships, postponement, supplier partnerships, and lean practices [39]. SCM comprises various steps of a product life cycle and usually requires the cooperation of several businesses and stakeholders. The various stages and the types of cooperators in the supply chain make it a deeply interconnected network that becomes challenging and cumbersome to maintain and monitor. Furthermore, SCM is examined not simply by the specifications on keeping of records but further by the elements incorporated among a particular business. Due to this, several consensus algorithms and blockchain frameworks have been advised to address concerns in distinct products, industries, and businesses [1]. Hence there are various surveys and innovations yet to be done based on different types of businesses and industries that motivate using blockchain as a viable option.

Various surveys in the food, beverages, and agriculture industry have proposed to adopt blockchain technology based on the Interplanetary File System (IPFS) and

Ethereum. IPFS is a distributed file-sharing method with fast data retrieval speeds that ensures easy and fast accessibility of data to numerous stakeholders of the supply chain.

4.1.1 Digitization of SCM

Digitalization is simply a process which when carried out correctly, can be heavily transformative for a business, but that is only possible when a business has successfully digitized first. It helps in converting products and information, such as records, ledgers, delivery details, payments, etc., into a digital format [40]. Once the conversion process is done, these resources can be used to streamline and upgrade processes, eliminating the need for basic information such as paper, meetings, and face-to-face interaction which is something that can be especially helpful in the hard times that COVID-19 has presented.

4.1.2 Smart Contract as Part of SCM

Smart contracts are simple programs stored on blockchain that run only when the predefined conditions are fulfilled. They help in automating the execution of an agreement so that participants are aware of certain outcomes without any wastage of time or involvement of any negotiator. This self-executing nature helps in reducing the complexities and challenges in operation of the supply chain. The smart contract being digitized protocols can concede the tracking of the supply chain, transactions, and documents that fulfill and streamlines the business processes, and boost the accomplishment of business logic [41]. Therefore, supply chain digitization is anticipated to improve the engagement of stakeholders in achieving similar objectives and goals.

4.2 Advantages of Leveraging Blockchain

Blockchain benefits organizations to understand their supply chain network and employ consumers with real-time, authenticated, and immutable data, also it aids supply chain management in various ways. Some technical advantages of using blockchain technology are classified based on the particular supply chain difficulty addressed by businesses.

4.2.1 Immutability

Blockchain provides the feature of immutability that accommodates originality, traceability, transparency and helps in the verification of records. This feature protects every entity (transaction records and details) from being replaced and manipulated inside a blockchain. For instance, the immutability feature protects reports concerning provenance, traceability, and file confirmations. Although, the system that promotes immutability without altering transactions is still developing.

4.2.2 Transparency

Transparency is an essential feature for any successful supply chain. As every segment of the supply chain network directly connects all others, it is vital to maintain clear communication channels for all parts to work cohesively. In terms of data transfer, each level of the system needs to receive the same report simultaneously. Numerous supply chain logistics businesses have achieved a certain level of this transparency by the use of blockchain. Blockchain provides all functions within a particular supply chain with access to identical information through a chain of data sharing. It is a technology that allows authenticated data to communicate between individuals without the need of a middle-man or conferred central party [42].

4.2.3 Security

The risk posed to a digital supply chain is slightly different from that typically present in those linked with physical goods. The significance of data, and other digital information, however, makes it a more lucrative target. The intruders target shipping manifests, invoices, tracking data, and reconstruct them in order to get access to the actual supplies. Vulnerabilities such as cargo thefts fictitious pickup are one of the major concerns for the disruption of supply chain operations that are running internationally. Hence blockchain helps in overcoming such issues, by providing encryption and validation. It ensures that your data is encrypted, meaning modifications in data cannot be performed. It also performs various cryptographic encryption to generate a cryptographic signature of a document or file. This gives users a gateway to ensure a file is un-tampered, without requiring them to save the entire data on the blockchain [43]. Due to its decentralized nature, you can always verify file signatures across every ledger on every node in the network and verify that they haven't been altered.

4.2.4 Streamlined Operations

Blockchain enables so many possibilities for supply chain like smart contracts, proof of provenance, transparency, real-time responses, and many more which helps companies to operate in an efficient way by streamlining processes like product development, manufacturing, quality control, shipment, or delivery, etc., all together. It supports industries and businesses by streamlining their operations processes to deliver their product and services to the consumers [44].

4.2.5 Quality Assurance

Blockchain helps in quality control by maintaining data transparency throughout the entire blockchain network. Quality control or assessment is the analysis of collected data (example: batch no. of a product, date, and place of origin, manufacturing date, etc.), through which the degree of assent with established standards and criteria is analyzed for a product or service. If the quality through this manner is found unsatisfactory, then investigations are conducted based on the parameters to identify the causes, and improvements are added to maintain the quality standards of the product according to the norm. If needed the control persons of the company or industry can reconstruct the critical issues from that particular stage in the process that caused a compromise in the quality indices in order to make the necessary corrections [45]. There are two important factors responsible for quality assessment which are, monitoring quality and analyzing anomalies.

4.2.6 Tracking Accuracy

In the supply chain industry, track accuracy refers to identifying the present and past locations of all inventory and products, as well as their history of custody. For record-keeping and tracking accuracy, a huge number of transactions is stored in blockchain. This huge data requires a lot of space which gives rise to various other challenges regarding the storage, that creates an electronic record of all kinds of data regarding the product and, vital in protecting consumers from exposure to stolen, contaminated, or counterfeit products. Tracking accuracy plays an important role in the food supply chain industry to identify the defective and decayed food products present inside the supply chain [46]. The size of distributed ledger is dependent on the number of transactions stored on it.

4.2.7 Cost Reduction

Blockchain can help in cost reduction in many ways, such as fostering trust within the organization, by automatically reducing the need for third parties and middlemen which will lead to limited use of resources and will eventually reduce cost. Also, as each member will be linked to a digital distributed ledger, the cost and effort spent on documentation will ultimately be reduced. Blockchain also helps in reducing the transaction cost that aids in the financial sector by replacing a digitized record-keeping infrastructure and substituting legacy systems. It also reduces the need for payment negotiators like money transferring services, stock exchanges, and payment networks. Many businesses are harnessing blockchain as it reduces cost in terms of the transaction including governance decisions allowing transparency and valid transactions throughout [47].

4.2.8 Provenance

Provenance means keeping of record of the data, this feature allows businesses to collect their data which is open to all and can also be verified using information available on the distributed ledger. It is one of the very explored features of blockchain in reported studies. It mainly resolves three challenges: tracing a product, tracking the origins of a commodity, and classifying defective products. Tracking the source of a product is critical if the need arises to suspend the defective or fake product quickly. For instance, we can take an example of the incident in 2015, when the restaurant chain suppliers Chipotle immediately retrieved their product in which spread E. Coli and salmonella. Detailed provenance data could have been more useful in such a scenario to provide a dynamic response.

Blockchain can help businesses and companies to classify the batch of products more accurately which can help in easy retrieval of contaminated items and can reduce the risk for users. It is an appropriate technology that can maintain a provenance record at various stages of a business, as the ownership of the commodity is changed, they are reflected inside the distributed ledger [48]. Figure 2 depicts the advantages of using blockchain technology when applied to supply chain management.

Blockchain technology can be linked to various important topics of Operations in Supply Chain Management (OSCM) exhibiting its potential impact in many fields such as agriculture industries, automobile industries, e-commerce, and in many other public and private sectors. Furthermore, there is only a limited no of applications based on OSCM literature in a blockchain, there will always remain a need for consensus encompassing the limitations and benefits around this technology. There is much more to explore in the field of OSCM and developing technology related to

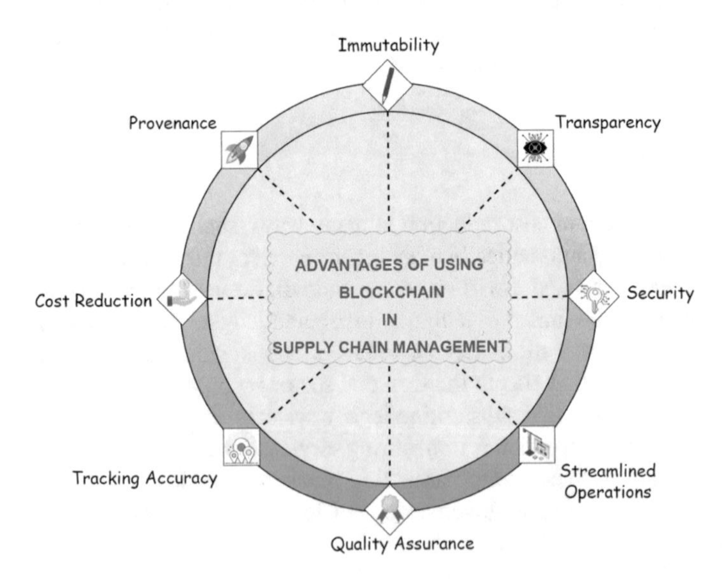

Fig. 2 Advantages of blockchain in supply chain management

the problems concerning OSCM, such as transparency, counterfeiting, sustainability, etc., that have also been explained in previous applications, blockchain additionally proposes critical issues concerning the importance of trust in supply chain and relevant studies regarding them [49].

4.3 Leveraging SCM in Industries with Blockchain Technology

The Blockchain data structure by default satisfies several requirements such as transparency, fraud prevention, fast settlement, scam-free, no third-party involvement enhanced security for many industries and organizations which help in the effective and efficient management of their operation, hence it is an undeniable option for industries and their partners to use this technology for managing their supply chain.

4.3.1 Benefits of LSCM

Blockchain databases are decentralized ledgers, by which the provenance of commodities can be determined when no individual participant can claim ownership regarding the overall data of the supply chain. Blockchain technology can help in maintaining and managing records of the date, location, price, certification, quality, and other relevant information of products and services to manage the supply chain more efficiently [50]. The availability of these records within the blockchain might improve traceability of the material inside the supply chain, also it can enhance an organization's position as a leading manufacturer by improving visibility and compliance over outsourced deal manufacturing, or by lowering losses from the counterfeit and gray market. SCM holds the combination of every business process across the entire chain of processes, and also it entails the engagement of all business functions that concern relationship management. The way of achieving strategic and business objectives by supply chain partners adopting this technology depends on the combination of industrial methods with the sort of blockchain used.

4.4 Use Cases of Blockchain in SCM

Blockchain is an emerging technology that is being used in many industries and businesses to achieve authenticity, transparency and to overcome some difficulties in their operation processes. Also, it adds an extra layer of security for them. Blockchain is now widely used in many businesses and industries, some of them are discussed here.

4.4.1 Food and Agriculture Supply Chain

Blockchain technology plays a major role in the domain of food and agriculture, taking the instance of the food and agriculture supply chain, those are the individuals, which draw more interest in implementing blockchain technology within the domain compared to other industries and businesses. Some of the addressed hurdles are quality control and assurance, transparency, performance improvement, provenance, and achieving sustainability. Food and agriculture supply chains involve many stages and it is not possible for them to be finely monitored and tracked which affects end consumers, and they are not able to trace their food or agricultural products source [51].

4.4.2 Diamond Production

Blockchain is playing a pivotal role in the diamond market. In the early 20 s, the journey of diamonds from mines to the customers was not easy to trace due to which many illegal activities took place. Later in 2015 many companies came up with the idea to use blockchain. A company named Everledger was founded to deliver transparency in the diamond business. With the help of IBM, Everledger made a blockchain with the help of open-source software called Hyperledger software. Due to which approximately 1.6 million diamonds are registered on this blockchain and sorted with 40 different metadata along with a high-quality image of diamonds in it. During every step of this manufacturing process, Everledger's process permits their customers to register data, like date, time, and other details including the signature of the person delivering it, this eases the diamond tracking as it moves through the supply chain [52]. Customers can simply log in with their credentials and view the entire provenance, which helps in reducing unauthorized activities by rogue company's and fraudsters and keeping them away from the bay.

4.4.3 Entertainment and Tourism

Blockchain helps various businesses like Integrated Casinos and Entertainment Logistics (ICEL) to utilize this technology by involving control of their businesses such as hotels, tourism, retails, etc. It is absolute for ICEL to assure that all members concerning the supply chain provide services and goods which serve those quality criteria established by industries and as per consumers' demands. Here blockchain technology is an ideal applicant which can resolve ICEL because it assures that every transaction inside the network is transparent and easy to trace, manage and identify [53, 54].

4.4.4 Vaccine Distribution

During the pandemic of the Covid-19, vaccines played a very important role in the survival of society and civilization. But the vaccine is useless if it is not delivered on time to the right person, this gap is fulfilled by blockchain which is helping in vaccine distribution. IBM created a distribution network of blockchain and supply chain solutions to help manufacturers proactively monitor for unpropitious events and enhance recall management [55]. It is an open-source blockchain platform to track safely and authenticate the lot and batch number level of vaccines, handling, temperature, and other histories. Distributors gain real-time monitoring and enhanced ability to counter supply chain disruptions.

4.4.5 Extermination of Slavery in Chocolate Industry

The western region of Africa, Ghana, and Ivory Coast are the mass producers of cocoa, which makes Africa the world's largest cocoa producer with 60% of the production of more than 3 million tons of cocoa but it is all done by slavery labor. Tony's Chocolonely a company that makes chocolate which is based in Amsterdam, renders a business that is helping to end modern slavery and child labor in this chocolate industry-based supply chain moreover to aid in creating a slavery-free chocolate industry. Tony's Chocolonely in association with Accenture, has developed and led a working prototype of a private blockchain, and its supply chain allies in the Ivory Coast strongly experimented with it in the field. The company engaged the cooperation of two of its associates in Ivory Coast, Socoopacdi, a farmer cooperative, along with Ocean a native tradesman, to try and examine the blockchain system, that includes a web app to input data, and integrate the service between applications, multichain blockchain, and the cloud infrastructure. Some particular associates were assigned with inputting data at three different steps, collection of the bean, buying of beans by the local trader, and local merchants selling beans to international traders. By this process, blockchain is helping in tracking each shipment of beans within the supply chain [56].

Table 3 presents an overview of the different use cases of blockchain technology used in supply chain management for enrichment of the economy.

5 Conclusion and Future Research Directions

The importance of data and security in SCM nowadays is very important which leads to the rise in security concerns. Many SCM-based industries are still following old orthodox methods which leave them vulnerable to different kinds of interventions. Therefore, it is important to ensure that the exchanges and the data remain safe, traceable, reliable, transparent, and secure. Hence in such a network, a distributed and decentralized technology can provide solutions to the issues. The rise of the

Table 3 Use cases of blockchain in SCM

Use cases	Description
Blockchain integration with food and agriculture supply chains [51]	• Solving problems like identifying counterfeit products by quality control • Maintaining transparency in the supply chain • Achieving sustainability at each level of operation
Blockchain for diamond production [52]	• IBM and Everledger collaboratively made a blockchain • Keep track of diamonds with 40 metadata along all the way to the consumer • Reduces unauthorized activity by rogue companies
Blockchain for entertainment and tourism [53, 54]	• Helping integrated casinos and entertainment logistics • Ensures the proper flow of goods by all members • Supports in maintaining the quality standards of hotels and casinos
Blockchain technology in vaccine distribution [55]	• Aids in proper distribution of vaccines at required locations • Ensures that the vaccine is delivered on time • Lets the distributors gain real-time access to overcome problems
Extermination of slavery using blockchain technology in chocolate industry [56]	• Ghana and Ivory coast are mass producers of cocoa beans • Child labor and Slave labor are used for production process • Tony's Chocolonely is a chocolate company helping in establishing cloud-based blockchain • This technology is helping them establish a transparent and slave free supply chain network

use of blockchain technology in the field of SCM has grown drastically, hence it is important to ensure that the exchanges and the data remain safe, transparent, and protected. Blockchain technology offers automation, utilizes smart contracts and authorized transactions moreover by default, it provides authentication, integrity, and transparency. In this paper, we performed a thorough survey on blockchain technology and its utilization in the field of SCM, in implementing enhanced security services such as integrity, privacy, confidentiality, provenance, and authentication support. We looked into the use cases and tried to identify the current supply chain-based security threats and investigated the security measures that are capable of decreasing such attacks. In the end, we discussed the advantage of using blockchain in the supply chain for various security purposes. For the future, we continue our effort to design and develop a decentralized application that eases the monitoring and tracking of products from source to destination which will also address the needs of the supply chain industry. We aim to achieve transparency and verifiability through our solution.

References

1. Shakhbulatov D, Medina J, Dong Z, Rojas-Cessa R (2020) How blockchain enhances supply chain management: a survey. IEEE Open J Comput Soc 1:230–249. https://doi.org/10.1109/ojcs.2020.3025313
2. Juma H, Shaalan K, Kamel I (2019) A survey on using blockchain in trade supply chain solutions. IEEE Access 7:184115–184132. https://doi.org/10.1109/access.2019.2960542
3. Omar IA, Jayaraman R, Salah K, Debe M, Omar M (2020) Enhancing vendor managed inventory supply chain operations using blockchain smart contracts. IEEE Access 8:182704–182719. https://doi.org/10.1109/access.2020.3028031
4. Chang SE, Chen Y (2020) When blockchain meets supply chain: a systematic literature review on current development and potential applications. IEEE Access 8:62478–62494. https://doi.org/10.1109/access.2020.2983601
5. Yang X, Li M, Yu H, Wang M, Xu D, Sun C (2021) A trusted blockchain-based traceability system for fruit and vegetable agricultural products. IEEE Access 9:36282–36293. https://doi.org/10.1109/access.2021.3062845
6. Haque AK, Bhushan B, Dhiman G (2021) Conceptualizing smart city applications: requirements, architecture, security issues, and emerging trends. Expert Syst. https://doi.org/10.1111/exsy.12753
7. Kumar A, Abhishek K, Bhushan B, Chakraborty C (2021) Secure access control for manufacturing sector with application of ethereum blockchain. Peer-to-Peer Netw Appl. https://doi.org/10.1007/s12083-021-01108-3
8. Wang S, Li D, Zhang Y, Chen J (2019) Smart contract-based product traceability system in the supply chain scenario. IEEE Access 7:115122–115133. https://doi.org/10.1109/access.2019.2935873
9. Hafid A, Hafid AS, Samih M (2019) New mathematical model to analyze security of sharding-based blockchain protocols. IEEE Access 7:185447–185457. https://doi.org/10.1109/access.2019.2961065
10. Belotti M, Bozic N, Pujolle G, Secci S (2019) A vademecum on blockchain technologies: when, which, and how. IEEE Commun Surv Tutor 21(4):3796–3838. https://doi.org/10.1109/comst.2019.2928178
11. Tsoulias K, Palaiokrassas G, Fragkos G, Litke A, Varvarigou TA (2020) A graph model based blockchain implementation for increasing performance and security in decentralized ledger systems. IEEE Access 8:130952–130965. https://doi.org/10.1109/access.2020.3006383
12. Asyrofi R, Zulfa N (2020) CLOUDITY: cloud supply chain framework design based on JUGO and blockchain. In: 2020 6th information technology international seminar (ITIS). https://doi.org/10.1109/itis50118.2020.9321013
13. Rouhani S, Deters R (2019) Security, performance, and applications of smart contracts: a systematic survey. IEEE Access 7:50759–50779. https://doi.org/10.1109/access.2019.2911031
14. Hader M, Elmhamedi A, Abouabdellah A (2020) Blockchain technology in supply chain management and loyalty programs: toward blockchain implementation in retail market. In: 2020 IEEE 13th international colloquium of logistics and supply chain management (LOGISTIQUE). https://doi.org/10.1109/logistiqua49782.2020.9353879
15. Reyes PM, Visich JK, Jaska P (2020) Managing the dynamics of new technologies in the global supply chain. IEEE Eng Manag Rev 48(1):156–162. https://doi.org/10.1109/emr.2020.2968889
16. Hassija V, Chamola V, Gupta V, Jain S, Guizani N (2021) A survey on supply chain security: application areas, security threats, and solution architectures. IEEE Internet Things J 8(8):6222–6246. https://doi.org/10.1109/jiot.2020.3025775

17. Fu Y, Zhu J (2019) Big production enterprise supply chain endogenous risk management based on blockchain. IEEE Access 7:15310–15319. https://doi.org/10.1109/access.2019.2895327
18. Musigmann B, von der Gracht H, Hartmann E (2020) Blockchain technology in logistics and supply chain management—A bibliometric literature review From 2016 to January 2020. IEEE Trans Eng Manag 67(4):988–1007. https://doi.org/10.1109/tem.2020.2980733
19. Wu Y, Yan Z, Thulasiram RK, Atiquzzaman M (2021) Guest editorial introduction to the special section on blockchain in future networks and vertical industries. IEEE Trans Netw Sci Eng 8(2):1117–1119. https://doi.org/10.1109/tnse.2021.3073636
20. Tran QN, Turnbull BP, Wu H-T, de Silva AJS, Kormusheva K, Hu J (2021) A survey on privacy-preserving blockchain systems (PPBS) and a novel PPBS-based framework for smart agriculture. IEEE Open J Comput Soc 2:72–84. https://doi.org/10.1109/ojcs.2021.3053032
21. Bencic FM, Skocir P, Zarko IP (2019) DL-tags: DLT and smart tags for decentralized, privacy-preserving, and verifiable supply chain management. IEEE Access 7:46198–46209. https://doi.org/10.1109/access.2019.2909170
22. Wan PK, Huang L, Holtskog H (2020) Blockchain-enabled information sharing within a supply chain: a systematic literature review. IEEE Access 8:49645–49656. https://doi.org/10.1109/access.2020.2980142
23. Bhushan B, Sahoo C, Sinha P, Khamparia A (2020) Unification of Blockchain and internet of things (BIoT): requirements, working model, challenges and future directions. Wirel Netw. https://doi.org/10.1007/s11276-020-02445-6
24. Saxena S, Bhushan B, Ahad MA (2021) Blockchain based solutions to secure iot: background, integration trends and a way forward. J Netw Comput Appl 103050. https://doi.org/10.1016/j.jnca.2021.103050
25. Bodkhe U, Tanwar S, Parekh K, Khanpara P, Tyagi S, Kumar N, Alazab M (2020) Blockchain for industry 4.0: a comprehensive review. IEEE Access 8:79764–79800. https://doi.org/10.1109/access.2020.2988579
26. Zhu P, Hu J, Zhang Y, Li X (2020) A blockchain based solution for medication anti-counterfeiting and traceability. IEEE Access 8:184256–184272. https://doi.org/10.1109/access.2020.3029196
27. Bernal Bernabe J, Canovas JL, Hernandez-Ramos JL, Torres Moreno R, Skarmeta A (2019) Privacy-preserving solutions for blockchain: review and challenges. IEEE Access 7:164908–164940. https://doi.org/10.1109/access.2019.2950872
28. Toyoda K, Machi K, Ohtake Y, Zhang AN (2020) Function-level bottleneck analysis of private proof-of-authority ethereum blockchain. IEEE Access 8:141611–141621. https://doi.org/10.1109/access.2020.3011876
29. Kim H, Kim S-H, Hwang JY, Seo C (2019) Efficient privacy-preserving machine learning for blockchain network. IEEE Access 7:136481–136495. https://doi.org/10.1109/access.2019.2940052
30. Sadiq A, Javed MU, Khalid R, Almogren A, Shafiq M, Javaid N (2021) Blockchain based data and energy trading in internet of electric vehicles. IEEE Access 9:7000–7020. https://doi.org/10.1109/access.2020.3048169
31. Kshetri N (2021) Blockchain and sustainable supply chain management in developing countries. Int J Inf Manag 60:102376. https://doi.org/10.1016/j.ijinfomgt.2021.102376
32. Bhushan B, Sinha P, Sagayam KM, Andrew J (2021) Untangling blockchain technology: a survey on state of the art, security threats, privacy services, applications and future research directions. Comput Electr Eng 90:106897. https://doi.org/10.1016/j.compeleceng.2020.106897
33. Dinh TT, Liu R, Zhang M, Chen G, Ooi BC, Wang J (2018) Untangling blockchain: a data processing view of blockchain systems. IEEE Trans Knowl Data Eng 30(7):1366–1385. https://doi.org/10.1109/tkde.2017.2781227
34. Bhushan B, Khamparia A, Sagayam KM, Sharma SK, Ahad MA, Debnath NC (2020) Blockchain for smart cities: a review of architectures, integration trends and future research directions. Sustain Cities Soc 61:102360. https://doi.org/10.1016/j.scs.2020.102360

35. Tangsen H, Li X, Ying X (2020) A Blockchain-based node selection algorithm in cognitive wireless networks. IEEE Access 8:207156–207166. https://doi.org/10.1109/access.2020.3038321
36. Wei Y, Xiao M, Yang N, Leng S (2020) Block mining or service providing: a profit optimizing game of the PoW-based miners. IEEE Access 8:134800–134816. https://doi.org/10.1109/access.2020.3010980
37. Kim T, Lee S, Kwon Y, Noh J, Kim S, Cho S (2020) SELCOM: selective compression scheme for lightweight nodes in blockchain system. IEEE Access 8:225613–225626. https://doi.org/10.1109/access.2020.3044991
38. Zhang J (2020) Deploying blockchain technology in the supply chain. Comput Secur Threats. https://doi.org/10.5772/intechopen.86530
39. Aslam J, Saleem A, Khan NT, Kim YB (2021) Factors influencing blockchain adoption in supply chain management practices: a study based on the oil industry. J Innov Knowl 6(2):124–134. https://doi.org/10.1016/j.jik.2021.01.002
40. Sharma PK, Kumar N, Park JH (2019) Blockchain-based distributed framework for automotive industry in a smart city. IEEE Trans Ind Inf 15(7):4197–4205. https://doi.org/10.1109/tii.2018.2887101
41. Abuhashim A, Tan CC (2020) Smart contract designs on blockchain applications. In: 2020 IEEE symposium on computers and communications (ISCC). https://doi.org/10.1109/iscc50000.2020.9219622
42. Montecchi M, Plangger K, West DC (2021) Supply chain transparency: a bibliometric review and research agenda. Int J Prod Econ 238:108152. https://doi.org/10.1016/j.ijpe.2021.108152
43. Wang L, Shen X, Li J, Shao J, Yang Y (2019) Cryptographic primitives in blockchains. J Netw Comput Appl 127:43–58. https://doi.org/10.1016/j.jnca.2018.11.003
44. Wong L-W, Leong L-Y, Hew J-J, Tan GW-H, Ooi K-B (2020) Time to seize the digital evolution: adoption of blockchain in operations and supply chain management among Malaysian SMEs. Int J Inf Manag 52:101997. https://doi.org/10.1016/j.ijinfomgt.2019.08.005
45. George RV, Harsh HO, Ray P, Babu AK (2019) Food quality traceability prototype for restaurants using blockchain and food quality data index. J Clean Prod 240:118021. https://doi.org/10.1016/j.jclepro.2019.118021
46. Köhler S, Pizzol M (2020) Technology assessment of blockchain-based technologies in the food supply chain. J Clean Prod 269:122193. https://doi.org/10.1016/j.jclepro.2020.122193
47. Schmidt CG, Wagner SM (2019) Blockchain and supply chain relations: a transaction cost theory perspective. J Purch Supply Manag 25(4):100552. https://doi.org/10.1016/j.pursup.2019.100552
48. Cui P, Dixon J, Guin U, Dimase D (2019) A blockchain-based framework for supply chain provenance. IEEE Access 7:157113–157125. https://doi.org/10.1109/access.2019.2949951
49. Wamba SF, Queiroz MM (2020) Blockchain in the operations and supply chain management: benefits, challenges and future research opportunities. Int J Inf Manag 52:102064. https://doi.org/10.1016/j.ijinfomgt.2019.102064
50. Perboli G, Musso S, Rosano M (2018) Blockchain in logistics and supply chain: a lean approach for designing real-world use cases. IEEE Access 6:62018–62028. https://doi.org/10.1109/access.2018.2875782
51. Bhutta MN, Ahmad M (2021) Secure identification, traceability and real-time tracking of agricultural food supply during transportation using internet of things. IEEE Access 9:65660–65675. https://doi.org/10.1109/access.2021.3076373
52. Thakker U, Patel R, Tanwar S, Kumar N, Song H (2021) Blockchain for diamond industry: opportunities and challenges. IEEE Internet Things J 8(11):8747–8773. https://doi.org/10.1109/jiot.2020.3047550
53. Yadav JK, Verma DC, Jangirala S, Srivastava SK (2021) An IAD type framework for blockchain enabled smart tourism ecosystem. J High Technol Manag Res 32(1):100404. https://doi.org/10.1016/j.hitech.2021.100404
54. Filimonau V, Naumova E (2020) The blockchain technology and the scope of its application in hospitality operations. Int J Hosp Manag 87:102383. https://doi.org/10.1016/j.ijhm.2019.102383

55. Ricci L, Maesa DD, Favenza A, Ferro E (2021) Blockchains for COVID-19 contact tracing and vaccine support: a systematic review. IEEE Access 9:37936–37950. https://doi.org/10.1109/access.2021.3063152
56. Musah S, Medeni TD, Soylu D (2019) Assessment of role of innovative technology through blockchain technology in Ghana's Cocoa beans food supply chains. In: 2019 3rd international symposium on multidisciplinary studies and innovative technologies (ISMSIT). https://doi.org/10.1109/ismsit.2019.8932936

"Implementation of Blockchain Technologies in Smart Cities, Opportunities and Challenges"

Ognjen Riđić, Tomislav Jukić, Goran Riđić, Jasmina Mangafić, Senad Bušatlić, and Jasenko Karamehić

Abstract A blockchain can be thought of as the shared distributed ledger type of technology that stores the information of every transaction in its network. The blockchain has emerged with vast diversity of applications in the economical and non-economical areas. Blockchain technology has the potential to provide a robust span of solutions to the issues faced in the implementation of smart cities. As such, it displays the potential to create smart types of contracts more securely, by eliminating the need for centralized authority. A blockchain can be envisioned as a secure decentralized database that stores information utilizing a peer-to-peer type of network. The blockchain can be seen as a type of special stack, where blocks could be placed or stacked on top of each other. Subsequent blocks composing the blockchain have to be linked to each other by cryptographic type of hash. In contemporary times an increasing interest in the concept of blockchain technology has been observed. This secondary research utilizes detailed literature review of multifaceted sources of information, such as peer-reviewed and quality academic journal articles from renowned databases. With the introduction of blockchain, numerous fields like banking, finance,

O. Riđić (✉) · S. Bušatlić
International University of Sarajevo (IUS), Hrasnička cesta 15, 71210 City of Ilidža-Sarajevo, BiH, Bosnia and Herzegovina

S. Bušatlić
e-mail: sbusatlic@ius.edu.ba

T. Jukić
University Josip Juraj Strossmayer, Trg Svetog Trojstva 3, 31000 City of Osijek, Republic of Croatia

G. Riđić
University of Economics for Management (HDWM), Oskar-Meixner-Straße 4-6, 68163 City of Mannheim, Federal Republic of Germany

J. Mangafić
Economic Faculty, University of Sarajevo (EF UNSA), Trg Alije Izetbegovića 1, 71000, BiH Sarajevo, Bosnia and Herzegovina

J. Karamehić
College Center for Business Studies (CEPS), Josipa bana Jelačića 18, 71250 Kiseljak, Bosnia and Herzegovina

S. S. Muthu (ed.), *Blockchain Technologies for Sustainability*,
Environmental Footprints and Eco-design of Products and Processes,
https://doi.org/10.1007/978-981-16-6301-7_4

healthcare, and supply chain shall experience positive effects. The sustainability of smart cities can be further enhanced and ensured with the application of blockchain technology. One major aspect wherein blockchain can play an essential role is real estate and smart cities. Blockchain and Smart Cities concepts are fated to influence the future of our planet in numerous ways. Incorporating blockchain into the expansion of Smart Cities will make it possible to have a cross-cutting platform that connects the cities' different services, adding greater transparency and security to all services and processes.

Keywords Blockchain technology · Smart cities · Opportunities · Challenges

1 Introduction

A blockchain, also known as a distributed ledger, represents a write-only data platform sustained through a large number of nodes that do not entirely divulge in one another. Numerous research studies bring together blockchain, image, and video processing algorithms. Some of these applications may entail actions against false videos, processing of medical images, encryption of images, followed by the management of the digital content rights. The outlined approaches enable making sure that video has not been tampered with in connection to time stamping, making it possible to be listed as proof in the court of law. Blockchain innovation's core segment is being entailed in an innovative set of procedures that enables information to be exchanged between different elements within a system. In that way, there are no intermediaries because members belonging to the system can be connected via identities, which are encrypted and with each other using distributed communication. Transactions are subsequently attached to a particular type of changeless ledger chain and distributed to each single node. With the increased incidences of information breaches, fraud, and extortion, numerous types of projects are utilizing blockchain innovation technologies in processing identity and document approvals. Validation by the means of blockchain is being enabled via timestamping, checking for legitimacy, and end-to-end encryption [9].

The blockchain can be conceptualized as a heap, in which the individual blocks are being stacked on top of the other. In this sequence, each subsequent block in blockchain is connected to each other by the special type of hash, called, cryptographic hash. The generated first block in blockchain is called the genesis block. These blocks are stored in the memory of the computers and run as a distinctive type of computer process. Knowing that each block is being constructed on top of the previous block, desired immutability is attained. The immutability implies inherent difficulty to fake/change a block and easiness to detect the tampering. Additional analogies for blockchain system can be visualized as the ledger book, whereby each block represents a separate page in the ledger, and each transaction represents an individual asset transfer on a ledger page. Each member in Blockchain includes nearly the same copy of the blockchain ledger. A blockchain transaction entails a transaction record in blockchain, similarly to a record warehoused in MySQL database. The blockchain network can either be in a government/public or private setting. In government run/public type blockchain, everyone is enabled to read or write transaction data with no need for various types of authorizations. In the private type of blockchain, no more than authorized nodes are enabled to read or write the transactional type of data inside the blockchain [4, 6].

2 Blockchain Enabling Technology Types

A blockchain imitates a central computing service by the means of a disseminated protocol, operated via nodes connected by the means of the Internet. In technical sense, the blockchain is supplanting the present integrated ledger structures with the decentralized types. A blockchain utilizes encoding systems, and it does not carry out the participation of a third part, thus making it stable and dependable. A blockchain is comprised of a distinct data blockchain. In this sense it is important to envision that building blocks can be constructed and scanned by particular participants. Its submissions are unchanging, clear, plain, and easy to use. Operations are documented sequentially in a constantly evolving databank. Computer structure is connected via the world wide web, whereby customers at every computer can obtain or disseminate information to different computers. In conclusion, the info is then being replicated and set aside via the structure by the means of a shared system. As such, it promotes mutual trade of substantial value devoid of a vital liaison [6, 7, 9, 14].

Blockchain technology is dependent on to extend significant capacity for effecting fundamental alterations in a wide range of company styles, operating processes, and enterprises by enhancing accounting and examining. Being in a kind of specific and multidimensional landscape, the necessity arises to recognizing the major alterations in the private, governmental, and business-related fields. The present point of interest relating to blockchain innovation is using it in execution and approval of shop trades. This implies the rationale signifying the fact that its progress was profoundly rubbed by the money-driven trades. Presently, the blockchain is proliferating via other economic markets [9].

Figure 1, displays the essential elements of a blockchain.

(1) Replicated ledger: As a part of the blockchain, information is not deposited at a main viewpoint. Blocks tend to be dispersed and reproduced between the nodes. Every node includes a copy of the completed record book. As such, a peril of informational loss is being eliminated.

(2) Cryptography: Information, being deposited in the blockchain, is being encoded via the robust encoding type of set of rules. Therefore, the dependability of the combined operations and files is being reinforced by the means of numeral autographs.

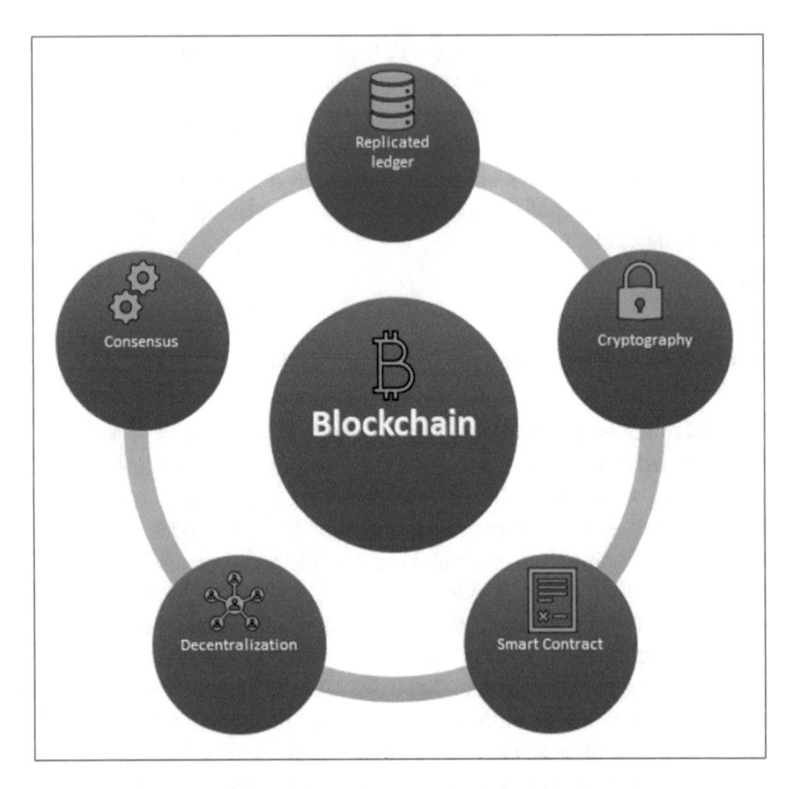

Fig. 1 Critical composing elements of Blockchain technology

(3) Consensus: Each building block includes a various total of operations. Operations must be authenticated prior to being added together to a standing building block. In Hyperledger type of structure, the validating nodes tend to be present to validate each block before attaching them into the chain.
(4) Decentralization: Aggregate transactions are distributed not including centralized type of control. Decentralization affords a sought after, safety and trust in the data, AND
(5) Smart contract: Blockchain delivers the electronic form of the contract connecting two sides. The smart contract can be visualized as the computer encryption aimed to numerically enable, authenticate, or impose various arbitration types [9].

Table 1 depicts the aggregate literature review of the Blockchain related research activities to be applied in the Smart cities. It starts chronologically from the year 2016 up until the year 2019, and it includes various Blockchain platforms (i.e., Ethereum, Bitcoin, Hyperledger Fabric, Multichain, Block-VN, and BigchainDB). It is being followed by data types (e.g., Medical history, user data, electricity, e-voting, supply chain, video, etc.). Finally, there are descriptions of its applications, such as patients' personal immutable medical record, secure electronic voting system,

Table 1 Summary of Literature review in regard to Blockchain related research articles to be applied in the smart cities

Year	Blockchain platform	Date type	Description
2016	Ethereum	Medical history	Patient's personal immutable medical record
2017	Bitcoin	User data	User-centric access control of personal data
2017	Hyperledger fabric	Anonymized dataset	Consensus based data transfer between data broker and data receiver
2017	Ethereum	Distributed energy resources	Distributed energy resources control system for smart grids
2017	Multichain	Electricity	Proof-of-concept based blockchain for electricity trading in smart industry
2017	Block-VN	Vehicular information	Distributed network of vehicles in smart city
2017	BigchainDB	Supply chain data	Storage of products data of food supply chain
2017	Hyperledger fabric	Video	Smart contract and network service based blockchain for video delivery
2018	Ethereum	E-voting	A secure electronic voting system based on blockchain
2019	Hyperledger fabric	Drug records	Integrity management of drug supply chain for smart hospitals
–	Hyperledger fabric	Video, metadata	Data verification system for CCTV surveillance camera for smart cities

Source [9]

integrity managing of prescription medications' supply chains for Smart hospitals and customer-centrical control of individual information) [9].

Blockchain technology initially was mentioned in a research publication by Haber and Stornetta, in 1991. To better comprehend blockchain technology it is essential to describe some of its elementary notions:

(1) **Nodes**: represent a most elementary Blockchain's component. The blockchain is being constructed utilizing a distinct web of nodes. In real life the nodes could be visualized to computers;
(2) **Transactions**: Every distinct segment in a Blockchain implies the particular operation. If there is a desire to alter a worth on the Blockchain a new transaction will have to be generated, transmitting the computer-generated type of paper money from single bank account to another (all of which constitutes another

type of transaction). In order for a particular operation to be recognized, it ought to be endorsed by at least 50%+1 of the present nodes;

(3) **Block**: shows a way as to how a blockchain retains the information. A building block entails the information from numerous, distinct, transactions. Each block is connected to the previous block by a cryptographic type of hash. The sum of these blocks are, in turn, being stored in each, particular node; AND

(4) **Account**: Blockchain accounts entail two distinctive variables, a private and a government key. The account owner is essentially the private key bearer. Contrary to previous centralized technologies, in Blockchain, in the scenario of the loss of the private key, the possibility to claim the account, does not exist [15].

3 Accompanying Features of the Blockchain Technology

Main features of Blockchain technology are being described, in detail, below:

(1) **Decentralization**—the data is being reposited in a number of locations, as numerous nodes are in the grid;

(2) **Scalability**—refers to the fact that there is an endless number of nodes in the grid;

(3) **Safety/Security**—utilizing a present expertise a Blockchain is in theory impossible to crack. As discussed earlier, in order for the business operation to be authorized 50% in addition to one additional node in the network are required to acknowledge it. In the event, whereby an attacker and potential intruder manages to change a blockchain or modify a particular single data portion, a new-found block is formed that is required to be authenticated by all the devices inside the blockchain network. 50% of data plus one node should be altered in order for illegal operations to be recognized and all of them must be broken down into/hacked simultaneously. If any single node is to react in a different way comparing to the others, the cryptographical hash linked node is checked, and node should be disregarded by the network while waiting for it to return to actual informational edition.

(4) **Intelligence**—past the elementary Blockchain technology it is still feasible to compose the custom code for every submission in a separate manner, thereby leaving space for several regulations and use instances; AND

(5) **Auditability**—since every block is connected to the previous one by the means of a hash, the Blockchain permits to circumnavigate through entirety of the blocks all the way up to the "Genesis" block. It represents the initial block of the Blockchain, thereby allowing for sequential tracing of all modifications [15].

Blockchain represents a regionalized P2P network utilizing the sum of produced operations are authenticated via the recorded nodes. These are then documented in a disseminated and unchangeable type of ledger. Therefore, the compromised

	Bitcoin	Hyperledger Fabric	Ethereum	Multichain	IOTA[a]	EOS.IO	Libra
Release year	2009	2017	2015	2015	2016	2018	2020
source	Open-source	Open-source	Open-source	Open-source	Open-source	Open-source	Open-source
Network Type	Public	Private	Public	Private	Public	Public	Public
Ledger type	Permissionless	Permissioned	Permissionless	Permissioned	Permissionless	Permissioned	Permissioned
Hashing algorithm	SHA-256	• SHAKE256 • SHA3	• Ethash • KECCAK-256	SHA-256	Curl-P-27	SHA-256	• SHA-3 • HKDF • Ed25519
Consensus algorithm	PoW	PBFT	• PoW • PoS (Serenity)	PoW	• PoW • Tangle	DPoS	LibraBFT[b]
e-currency	bitcoin (BTC)	N/A	Ether (ETH)	N/A [c]	IOTA	EOS	Libra
TPS	7	3500	15-20	200-1000	500-800	4000	1000
Smart contracts	Bitcoin Script	Chain-code	Smart contract	Smart Filters [d]	Not supported[e]	Smart contract	Move composed Smart contract

Fig. 2 List of contemporary blockchain platforms with source, network, algorithm, and ledger types. *Source* Majeed et al. [17]

algorithm signifies the hearth of blockchain technology. As such, it ensures the vital dependability of the network. More precisely, since no fundamental authority is present to confirm the created occurrences, each operation must be authenticated by the blockchain nodes by the means of a joint arrangement (i.e., accord). In the discussion, below, the elements of the most prevalent consensus kinds are being listed and discussed:

(1) **Proof-of-Work (PoW)**—a transaction is being authorized when at least 50% plus one of the nodes accept it in the P2P network,
(2) **Proof-of-Stake (PoS)**—each particular node that entails more capital has impacts bigger likelihood to partake in the consent and make a block.
(3) **Proof-of-Importance (PoI)**—the nodes able to build the block impact the largest number of transactions in the network. AND
(4) **Proof-of-Authority (PoA)**—only particular nodes are unequivocally permitted to establish new blocks and fortify the blockchain [1].

Figure 2, listed, above, represents a list of contemporary Blockchain platforms including source, network, algorithm, and ledger types. It compares and contrasts various Blockchain platforms (i.e. Bitcoin, Ethereum, Hyperledger fabric, Multichain, IOTA, EOS.IO, and Libra) with release years (2009–2020), types of sources (open versus closed), hashing algorithms, etc.

4 Smart City—The Origins

The starting point relating to the smart city comes from the progress in the quality of life of citizens and optimum supply operation of the city, due to the modern increase in speed in urban living. The progress in public services and infrastructure has improved the quality of life. These improvements were achievable due to the world wide web (WWW) and Internet, communication, and information technology improvements.

Special prospects stemming from the idea of smart cities include variety of efficient and effective public services, augmented by improved infrastructure, all of which are being easily accessible and more interactive. Vision of smart city became a reality with the potential of the Internet of things (IoT) concept.

Consequently, the smart city emerged as one of the primary generators in IoT applications. The complete city is thereby protected with the physical items, which are, in turn, intertwined with the IoT scheme. The four building blocks that can be joined together utilizing IoT concepts are (1) data, (2) phenomena, (3) people, and (4) processes. The Internet of Everything (IoE) came out including the people in the IoT paradigm, where an interlocked network is grouped in IoE. In conclusion, the image of a smart city is incorporated with IoE foundation blocks to enable encouraging services in the future [16].

5 Smart Cities—The Concept

Smart Cities can be visualized as ecosystems that are usually characterized as networks of connecting appliances. Their environs are customarily portrayed as puzzling systems created within a prism of supply mutuality. Gretzel et al. added four progressively essential elements that exist in this ecosystem's characterization of a smart city. These are (1) self-organization, (2) interaction or engagement, (3) balance, and (4) lightly combined performers with common purposes [9].

The Smart City notion is characterized in the literature in various ways. Researcher Komninos depicts smart cities and associated areas as environments with a great capacity for knowledge creation, expansion, and innovation. They are including the ingenuity of population and institutions with digital infrastructures to operate in the physical, institutional, and digital spaces of cities. The opacity of this notion triggers problems in comprehending the way as to how information technology influences the development of smart cities [15].

6 The Smart City—The Model

The high-level point of view of the smart city paradigm is depicted in Fig. 2. It displays how the differint components in a smart city convey to each other in order to deliver the services in smart cities. Various sorts of elements are part of smart cities. These elements interrelate with each other by the means of cellular and/or Internet services (e.g., ZigBee, Wi-Fi, 3G/4G/5G/6G). The smart energy, smart mobility, and smart grid, and various services in the smart city surroundings. The macro or microcell entities represent a type of communication gadgets utilized to make available services on demand [16].

As depicted in Fig. 3, the Smart city model displays critical connections and interactions between Smart City with Fog computing and internet of things (IoT)

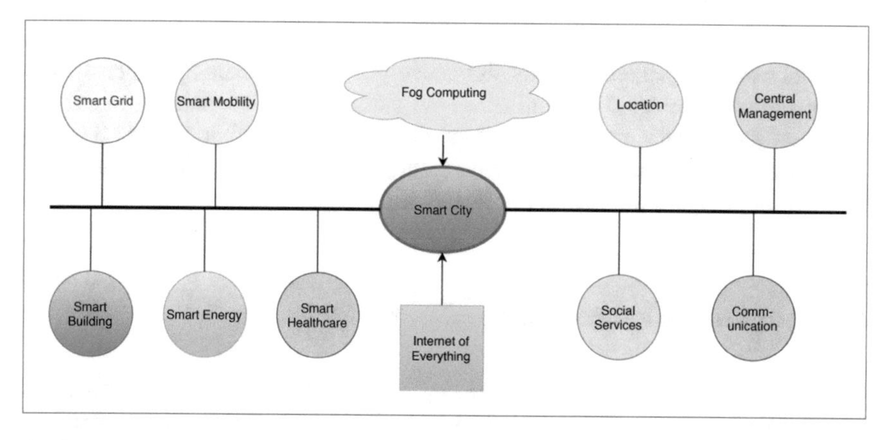

Fig. 3 The model of the smart city. *Source* Singh et al. [16]

as the main building blocks. Additional contributing elements are being exemplified in smart grid(s), smart buildings, smart energy, smart mobility, social services, communication, location, capital management, and smart healthcare [16].

7 Primary Objectives of the Smart Cities

In order to make certain that the advantages associated with urban growth are fully utilized, urbanization management policies ought to enable access to the electronic infrastructure and services for the entirety of its residents. The attainment of confidential data through the Internet of Things (IoT) represents one of the major objectives of smart cities. Therefore, the establishment of this data security is of utmost importance. Therefore, most administrations invest regularly in the development of smart cities with many designated facilities for its inhabitants. The modern city already represents a robust network of interconnected technologies, and according to Cisco technologies, 500 billion devices are expected to be connected to the Internet by the year 2030. In this context, the researcher Gartner stated that about 10 billion interconnected objects are expected to be utilized by smart cities by 2020 [15].

8 Smart City as a Summation of Paradigms

Smart City represents a sum of paradigms spread across various realms, such as people, economy, government, mobility, environment, and life on our planet earth. As such, it is inherently designed to address a range of utilization instances, such as: (1) environmental monitoring, (2) analysis of the traffic, (3) utility monitoring, (3) smart public transportation, (4) electronic voting system, (5) e-commerce, (6) jobs, (7) local occurrences, (8) real-time incident reporting, (9) medical services, etc. Data analysis gathered among the above-mentioned spheres permits the city administration to enhance the infrastructure and adjust its services. A smart city additionally represents, rather unique type of environment with integrated information and communication technologies creating interactive spaces that bring along computational capabilities to the physical world [15].

9 Composing Elements of the Smart Cities

A Smart City is designed to incorporate key elements that allow data centralization, elements that can take many shapes, starting from a simple website to complex applications, backed by specialized hardware. The accessibility of the data ought to be guaranteed in a way that the system can be freely accessed by citizens, enabling them to propose changes and corrections in an interactive way [15].

Numerous IoT devices require memory and computational complexities to deal with modern computing gadgets. Shortage of computational power makes them defenseless against a broad scope of cyber-attacks. They addressed the problem of security issues in relation to distributed refusal of service (DDOS) attacks in the IoT system. They used a changed smart contract that empowers a superior resistance mechanism against DDoS and rogue device assaults. Kim et al. introduced an idea of utilizing blockchain technology to address and resolve the security problems of a sensor-based platform. IoT gadgets represent the main components of smart homes, smart factories, and intelligent appliances. A blockchain-sourced authentication protocol was offered to focus on security problems. By utilizing that protocol, the IoT environment can become efficient and stable. They utilized this feature of blockchains in the IoT environment to safeguard the verification at runtime of sensors and actuators. The utilization of smart contracts makes it easy to automate the business logic and assists in saving time with the guarantee of zero error security. Significant volume of work has been performed in the area of video forensics. This improvement allows the video proof to be utilized in court cases. Recent techniques utilized for video falsification uncovering are primarily based on an autoencoder with periodic convolutional neural networks, augmented by an autoencoder with a go turn algorithm, watermarking techniques, and digital signatures. It was further proposed to include autoencoders and intermittent neural networks-based architecture to discover the video falsification. They produced a unique content-based signature to detect inter and intraframe falsified videos. In extension, irrespective of whether an image examination innovation is produced, there is a multitude of cases of breakdown due to the sensor-based restriction. This may similarly occur when restricting images for injurious aims. The Privacy Act takes into consideration the creation of CCTVs for public places, which requires reaching out to the owners of CCTVs to acquire video information. However, this represents a rather lengthy procedure. Irrespective of whether a video is acquired, it is difficult to utilize in open organizations due to the fact that video is not safeguarded to be an original video that has not been manipulated. Panwar et al. proposed an arrangement to provide sensor information confirmation through cryptographic algorithms implementing a log sealing system and creating permanent portions of evidence used for log verification. The structure ensures that sensor information and log-fixed data could be put away in untrusted storage with the proposed verification system ensuring its integrity. However, this structure depends on the dependability of the instrument; for example, Intel SGX store up the fixed data in an incorporated way. False news became a worldwide issue that raises extreme difficulties for human culture and the majority rules system [9].

10 Application of the Blockchain Technology in the Smart Cities

From its inception, Blockchain technology was perceived through its relationship with bitcoin. Recently, its possible utilization has been investigated in other fields of activity. These may include smart contracts, logistics, and systems' management. Utilizing the promise of Blockchain, researchers and developers are aiming to increase people's trust in digital communities. This can be easily accomplished by Blockchain systems through their decentralized and open nature. Provision of a single source of truth and a single starting point for new initiatives represent some of the examples. Domains and applications of using blockchain technology in cybersecurity are presented [15].

These are shown in Fig. 4.

Figure 4, depicts the interaction between Blockchain domains and associated applications in Smart cities. Cybersecurity influences the need for the existence of Blockchain. Blockchain furthermore branches to distinct types of applications, process models, and communications infrastructure(s). In the end, the applications may involve, financial system, intelligent transportation system, IoT, Smart grid, data center networking, voting systems, and healthcare networks [15].

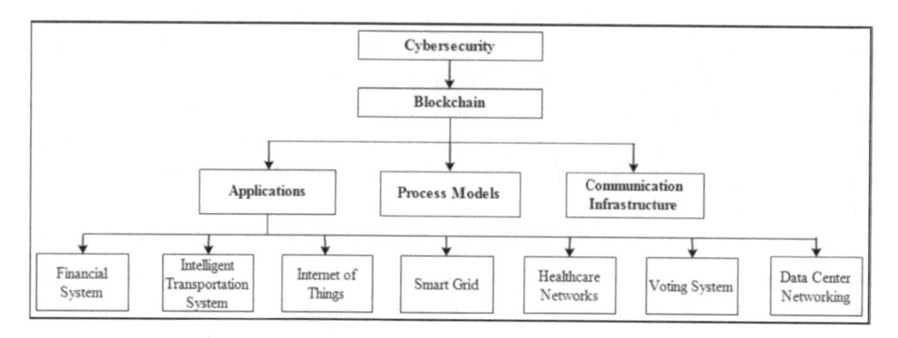

Fig. 4 Domains and applications of blockchain technology in smart cities. *Source* Rotună [15]

Blockchain technology is capable to augment the openness of local and regional institutions. It does so while enabling the communication of sensitive data without compromising security and confidentiality. In this sense, blockchain may be utilized in the advancement of smart cities. This can be achieved by utilizing an interoperable platform enabling the citizens to proactively participate in the decision-making processes. These processes affect the communities they belong to. They may also operate as a tool for managing the reputation of companies in relation to the activity related to the environment. The administration of a smart city via its systems generates a considerable volume of sensitive data that requires an oversized storage environment. According to latest data, cyber-attacks still pose a real security concern in realm of the online transactions. To mitigate the effects of these phenomena, blockchain technology utilizes a distributed model that increases the degree of entropy. This is being achieved by implicitly reducing the vulnerability of the systems it supports. The technology-based cryptography architecture makes it unlikely that transactions will be reversed or altered. Whenever a new transaction is broadcast on the network, the nodes have the obligation to validate and include it in the copy of the distributed ledger. In an invalidation scenario, it must simply ignore it. The consensus is reached when most of the participants composing the network decide on a single state. Furthermore, all participants in the system possess a personal key or signature utilized during creation of a transaction. This key allows the association between the user who created a particular transaction and the recipient of that transaction. At the same time, since the ledger is distributed and validated by the entire network, a transaction is associated with a single user and cannot be registered multiple times on the blockchain [15].

11 Application of Blockchain Technology in the Development of the Smart Cities

A smart city provides its inhabitants with the facility to interact with public administration and local communities, utilizing digital technologies for increased efficacy and safety. Through its persistent and at the same time distributed storage, blockchain permits the development of a large number of new interaction models. It is important to note that these models could not be designed within a centralized model. Public administrations are starting to realize the potential of the blockchain model as a platform for communication and transactions in the implementation of electronic services for local communities [15].

Figure 5, discusses the critical interaction between Smart city's citizens in connection to the primary circle involving computational intelligence, blockchain, IoT, etc. Secondary circle involves mobility, industry and services, governance, and health care. Finally, the third circle involves air quality monitoring, internet of value (IoV), traffic control, various sensors, information transparency, shared medical data, etc. [3, 13, 17].

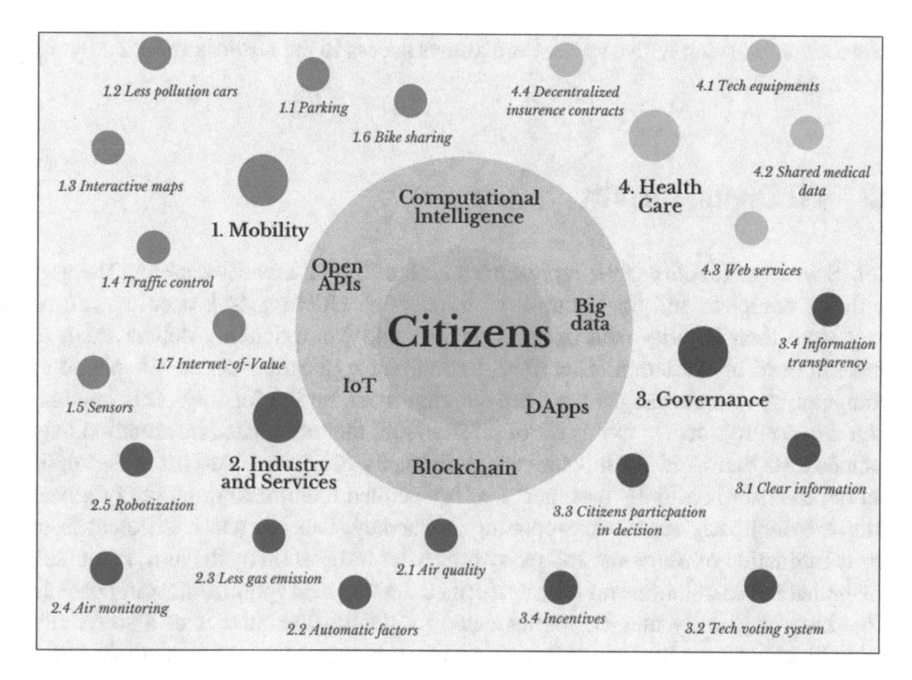

Fig. 5 The scope of smart cities with blockchain as potential trend. *Source* Oliveira et al. [13]

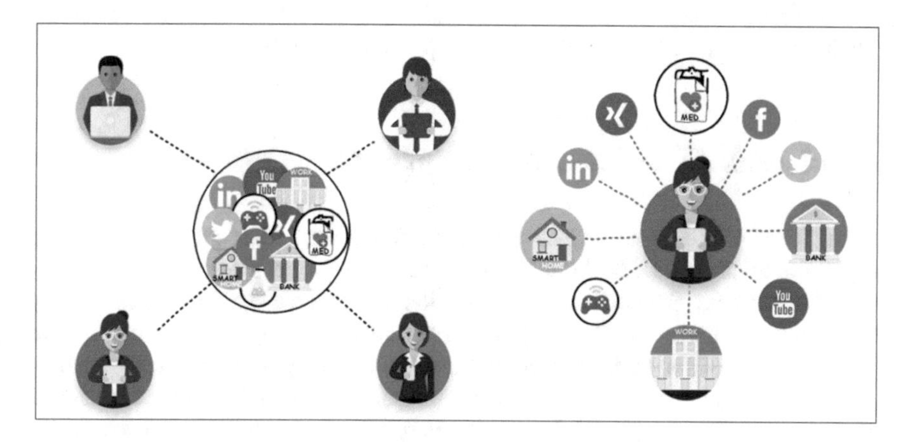

Fig. 6 The classical compared to SSI type of digital identification. *Source* Rotună et al. [15]

12 Digital Identity

Digital identity represents the information regarding an entity. It is utilized by the information systems to embody an external agent, in a form of a person, an organization, an application, or a device. ISO 24760-1 classifies this identity as an "entity-associated set of attributes". Digital identity data permits the automatic verification of a user networking with a system and allows access to the services supplied by the system [15].

13 SSI Digital Identity

Self-Sovereign Identity (SSI) represents a kind of digital identity enabling the user with the complete and final control of its identity. Utilizing SSI, users and firms may store their identity data on their devices and can efficiently deliver them to those in need of validation. Therefore, through an application, on mobile phone or computer, the user manages the elements that make up the identity. This application also controls access to this set of information. Identity related information may include birth dates, citizenship, university diplomas, or licenses. As part of the application, the user is initially assigned a self-generated identification number derived from the public key and a corresponding private key. This key pair is different from the combination of username and password. After its creation by the user, automatic mathematical calculations must be performed making decryption almost impossible This kind of identity may be implemented to identify the citizens of a smart city using blockchain technology. All of this ensures storage, secure timestamping, and decentralized hosting. This model removes the need for passwords and guarantees authentication with a high degree of security. An example of successful implementation is the Estonian e-Residency program allowing clients outside the EU to create a digital identity that can be utilized to set up a business in Estonia [15].

14 Blockchain Architecture for Smart Cities

Smart cities utilize various technologies and infrastructure to ensure a better quality of life for urban residents. In addition, there is also a need for a good environment for business development, optimization of resource use, and transparency for public administration. These goals may be achieved by utilizing blockchain working as a tool for decentralized and distributed ecosystems. Features, such as safety and transparency, shared information, common updating of the database, and information validation, provided by blockchain technology, empowers all smart city customers. Blockchain technology permits the interactions between citizens and local governments without the need for a central authority. Smart contracts optimize the functioning of the smart community through their ability to automatically execute transactions without the intervention of an operator [10, 15].

A blockchain-based Smart City model using SSI is illustrated in Fig. 7.

Blockchain infrastructure links the local community with public administration. The admission to the ledger is permitted to all community members. Every member possesses its own synchronized copy of the common ledger. Furthermore, each participant has a Digital Self-Sovereign Identity that is utilized to authenticate the person

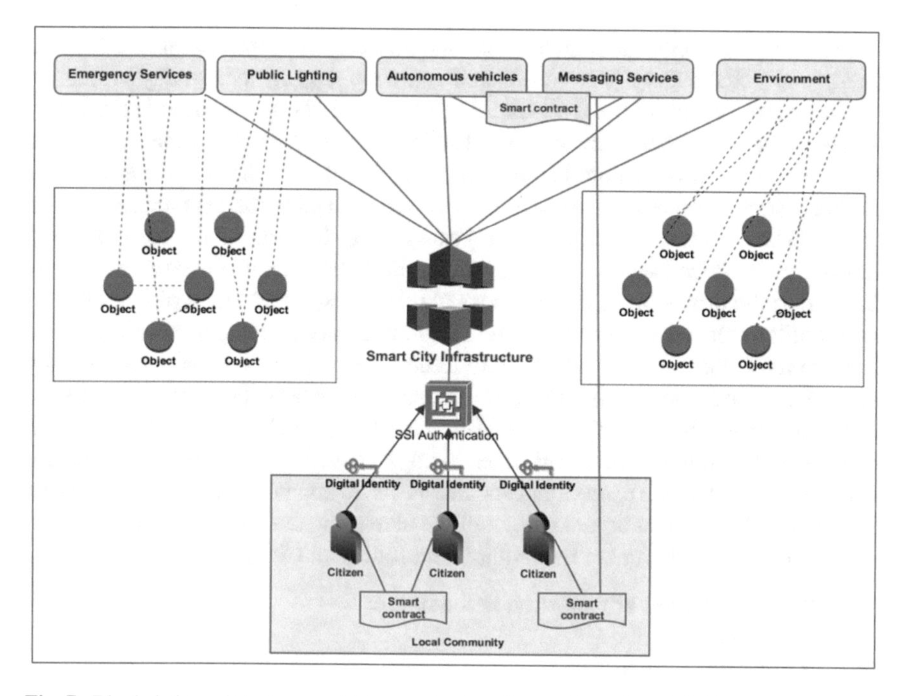

Fig. 7 Blockchain architecture model in smart cities. *Source* Rotună et al. [15]

in transactions. The utilization of central authorities is supplanted by a community of peers in the form of an interconnected network where each peer has its own identity. The ecosystem comprises IoT devices located in various locations, which record and transmit securely real-time data regarding the city environment. Various types of Smart contracts may be classified between citizens, between public authorities, and between citizens and public authorities for services. Smart contracts are stored on the blockchain thus decreasing the potential for fraud attempts [15].

15 Opportunities in the Application of the Blockchain Technology

The technology has played a main part when it comes to the advancement of cities. Stemming from first human settlements to the massive impacts of the industrial revolution to today, technological advancements have impacted our lives in a variety of ways. Experts envisage that the way we live today will change dramatically within the next decades. Big demographic and social change, followed by climate change shall impact our lives in the future and may impose an extreme burden on our existing city infrastructures [12].

Providing a sufficient amount of goods and services for people and huge cities will be especially challenging. Many cities won´t have the capacity for so many people, for instance in many parts of the world the electric grid is not made to provide electricity for such a huge number of inhabitants. The infrastructure must be renewed, and an interdisciplinary way of thinking will be needed. All over the world futurists are working on plans on how to make future cities more livable for such a massive gain in population. These imperatives have given rise to the birth of the idea of Smart Cities. Smart Cities is the idea of improving the efficiency of resource usage and facilitating the provision of new services within cities; by adding a digital layer to metropolitan area infrastructures through technology in an intelligent way in order to improve inhabitants' quality of life. Furthermore, Smart Cities aims to save money by reducing costs and also improving sustainability. Smart Cities' purpose is to use technology to enhance livability. The plan to create Smart Cities is not an easy plan. It needs to integrate communication technology as well as all sorts of information technology, together with important social aspects, to augment the standard of living considering the fact that urbanization will be increasing drastically in future [12].

The three major drivers of the requirement for Smart Cities are:

(1) Anticipated rapid urbanization in the future;
(2) Climate change; and
(3) Logistical pressures.

There is a strong impression that without addressing these three challenges, the world's mega-cities are predestined to become increasingly chaotic, inefficient, and susceptible to become increasingly controlled by the criminal elements [2, 4].

The scenario where Blockchain technology can have major impacts on Smart Cities is in the area of crime prevention, more specifically, the crimes involving forgery and counterfeiting may be prevented from alteration through the use of Distributed Ledger Technology (DLT).

As stated above, it can be concluded that the three principal drivers of the requirement for Smart Cities are: (1) Anticipated rapid urbanization over the coming decades; (2) Climate change; and (3) the Logistical demands. The notion that without addressing these three issues, the world's mega-cities are destined to become increasingly chaotic, inefficient, and susceptible to being controlled by criminal elements [4, 8, 16, 18].

The ongoing development of the Internet of Things (IoT)-based applications are paving the way towards the development of smart cities. Smart cities offer intelligent transportation, industry 4.0, smart healthcare, smart homes, smart banking, among others. These applications require immense security for handling data while improving the standard of citizens' life. To enable smart cities with enhanced security and privacy, we can use blockchain. Blockchain is a decentralized, traceable, transparent, and immutable ledger of transnational records in Peer-to-Peer (P2P) networks. Blockchain was first introduced as bitcoin that is a solution to transfer digital payments between different parties without the need for a central authority [3, 4, 17].

Other than improving the financial industry, blockchain has potential applications in many other fields such as the IoT, e-Commerce, accounting & auditing, e-Voting, asset management, identity management, supply chain, taxation, telecommunication, healthcare, and government/public services. The smart city comprises the ecosystem of smart environments provided in the city which can improvise its inhabitants' lifestyle. Smart city concerns with the adoption of information and communication technologies for enhancement in public welfare, economy, government services, environment, resource management, and urban planning. Smart cities envision the use of existing and developing digital technology to enhance every aspect of city life. One of the primary objectives of smart cities is reformed provision of fundamental services like housing, education, healthcare, transportation, energy, water, utilities, surveillance, and law enforcement. Smart cities mitigate the problems of population growth and expeditious urbanization by integrating social, business, and physical infrastructure of the city through technology. Recent advancements of technologies such as Information & Communication Technologies (ICT), blockchain, Big Data, machine learning, automation, Artificial Intelligence (AI), and the IoT will make smart cities more interconnected, instrumented, intelligent, livable, safer, sustainable, and resilient [5, 8, 9, 18].

The performance measures for the success of a smart city constitute the integration of fundamental services with seamless assimilation in the daily lives of its residents, thereby assuring the effective usage of resources and improving quality of life. However, this involves a huge amount of data traffic generated by information systems flowing through communication networks of city technological infrastructure. Blockchain is a solution to the key challenge of security, privacy, and transparency of this personal, organizational, and operational data [1].

16 Conclusion

A blockchain, also known as a distributed ledger, represents a write-only data platform sustained through a large number of nodes that do not entirely divulge in one another. Numerous research studies bring together blockchain, image, and video processing algorithms. Some of these applications may entail actions against false videos, processing of medical images, encryption of images, followed by the management of the digital content rights.

The blockchain can be conceptualized as a heap, in which the individual blocks are being stacked on top of the other. In this sequence, each subsequent block in blockchain is connected to each other by the special type of hash, called, cryptographic hash. The generated first block in blockchain is called the genesis block. These blocks are stored in the memory of the computers and run as a distinctive type of computer process. Knowing that each block is being constructed on top of the previous block, desired immutability is attained. The immutability implies inherent difficulty to fake/change a block and easiness to detect the tampering. Additional analogies for blockchain system can be visualized as the ledger book, whereby each block represents a separate page in the ledger, and each transaction represents an individual asset transfer on a ledger page. Each member in Blockchain includes nearly the same copy of the blockchain ledger. A blockchain transaction entails a transaction record in blockchain, similarly to a record warehoused in MySQL database. The blockchain network can either be in a government/public or private setting. In government run/public type blockchain, everyone is enabled to read or write transaction data with no need for various types of authorizations. In the private type of blockchain, no more than authorized nodes are enabled to read or write the transactional type of data inside the blockchain.

Blockchain-centered resolution for smart property provides a number of benefits over traditional centralized database systems. The first and most important advantage is the security against forgery. Since the blockchain depends on endorsement of transactions performed by verification of identity of all the parties in the network, false transactions cannot be verified without the authorization of all nodes in the network. This feature shall instantly resolve variety of the malicious issues faced stemming from identity stealing and fake payment plans. Transparency poses an additional significant advantage of the blockchain in contrast to traditional centralized databases. Since the blockchain is a decentralized shared database that is managed and coordinated across a network of devices (nodes), the data of every transaction remain constant among all the nodes. Therefore, numerous parties may utilize the exact same copy of data simultaneously. This is in contrast to centralized system where multiple nodes rely on siloed databases. This signifies improving the scope of transparency in the decentralized system. Another significant advantage of taking into consideration the blockchain as a solution for smart property systems is the efficiency. In theory, managing multiple copies of transactions in decentralized shared database in the blockchain could be considered inefficient in comparison to the siloed centralized database. In the real world, the organizations retain the copy of database including similar transaction data in situations such as data inconsistency, demanding expensive and time-consuming data resolutions. Consequently, utilizing a decentralized shared database solution such as blockchain may reduce the requisites for manual data resolution. As a consequence, a smart city cannot operate successfully if there is a lack of trust between organizations and gadgets comprising the network. Trust is the quintessence of combining all the components of a smart city together. The blockchain has proven itself to be a more efficient technology, as it offers framework for consensus in a decentralized ecosystem and provides complete transparency and immutability of information, thereby creating the blockchain as an essential layer of trust in the smart city. Many smart cities' exhaustive blockchain-centered decentralized solutions are being developed. As these platforms are expected to develop and mature, smart cities shall sooner or later function on a network of established blockchain solutions.

Smart cities have been increasingly becoming a reality, and their advanced services towards citizens often rely on IoT devices. Unfortunately, these IoT devices, are frequently poorly secured, leading to an optimal playground for cybercriminals, constituting a non-neglectable risk for the wide deployment and success of Smart cities.

References

1. Botello JV, Mesa AP, Rodríguez FA, Díaz-López D, Nespoli P, Mármol FG (2020) Block-SIEM: protecting smart city services through a blockchain-based and distributed SIEM. Sensors (Basel, Switzerland) 20(16). https://doi.org/10.3390/s20164636
2. Čirić Z, Ivanišević S (2019) Implementation of blockchain technology in the smart city. In: Conference: 5th international scientific conference on knowledge based sustainable development—ERAZ 2019At. Budapest, Hungary
3. Davidson S, De Filippi P, Potts J (2016) Economics of blockchain. https://hal.archives-ouvertes.fr/hal-01382002/document/. Accessed 20 Jan 2019
4. Dewan S, Singh L (2020) Use of blockchain in designing smart city. Smart Sustain Built Environ 9(4):695–709. https://doi.org/10.1108/SASBE-06-2019-0078
5. Durneva P, Cousins K, Chen M (2020) The current state of research, challenges, and future research directions of blockchain technology in patient care: systematic review. J Med Internet Res 22(7):e18619. https://doi.org/10.2196/18619
6. Ehrenberg AJ, King JL (2020) Blockchain in context. Inf Syst Front 22(1):29–35. https://doi.org/10.1007/s10796-019-09946-6
7. Grover P, Kar AK, Janssen M (2019) Diffusion of blockchain technology: insights from academic literature and social media analytics. J Enterp Inf Manag 32(5):735–757. https://doi.org/10.1108/JEIM-06-2018-0132
8. Hoxha V, Sadiku S (2019) Study of factors influencing the decision to adopt the blockchain technology in real estate transactions in Kosovo. Prop Manag 37(5):684–700. https://doi.org/10.1108/PM-01-2019-0002
9. Khan P, Byun Y-C, Park N (2020) A data verification system for CCTV surveillance cameras using blockchain technology in smart cities. Electronics 9(3):484. MDPI AG. https://doi.org/10.3390/electronics9030484
10. Kishigami J, Fujimura S, Watanabe H, Nakadaira A, Akutsu A (2015) The blockchain-based digital content distribution system. In: IEEE fifth international conference on big data and cloud computing, Dalian, pp 187–190
11. Mackey TK, Kuo T-T, Gummadi B, Clauson KA, Church G, Grishin D, Obbad K, Barkovich R, Palombini M (2019) "Fit-for-purpose?"—Challenges and opportunities for applications of blockchain technology in the future of healthcare. BMC Med 17(1):68. https://doi.org/10.1186/s12916-019-1296-7
12. Manushaqa L, Grant M, Baliakas P, Holotescu T, Amellal J (2019) Blockchain implementation in smart cities
13. Oliveira TA, Oliver M, Ramalhinho H (2020) Challenges for connecting citizens and smart cities: ICT, E-governance and blockchain. Sustainability 12(7):2926. MDPI AG. https://doi.org/10.3390/su12072926
14. Rodríguez B, Bolívar R, Scholl HJ (2019) Mapping potential impact areas of blockchain use in the public sector. Inf Polity: Int J Govern Democr Inf Age 24(4):359–378. https://doi.org/10.3233/IP-190184
15. Rotuna C, Gheorgita A, Zamfirou A, Smada DM (2019) Smart city ecosystem using blockchain technology. Inform Econ 23(4):41–50. https://doi.org/10.12948/issn14531305/23.4.2019.04
16. Singh P, Nayyar A, Kaur A, Ghosh U (2020) Blockchain and fog based architecture for internet of everything in smart cities. Future Internet 12(4): 61. MDPI AG. https://doi.org/10.3390/fi12040061
17. Swan M (2015) Blockchain: blueprint for a new economy. O'Reilly Media, Sebastopol, CA
18. Veuger J (2018) Trust in a viable real estate economy with disruption and blockchain. Facilities, 36(1/2):103–120. https://doi.org/10.1108/F-11-2017-0106

Blockchain Technology Enables Healthcare Data Management and Accessibility

Omar Ali, Ashraf Jaradat, Mustafa Ally, and Sareh Rotabi

Abstract Blockchain can be described as an immutable ledger, logging data entries in a decentralized manner. It has been argued that this emerging technology disrupts a wide variety of data-driven domains, including the health domain. In terms of data integrity, immutability, audit, data provenance, flexible access, confidence, privacy, and security, today's healthcare data management systems face critical challenges. A significant proportion of current healthcare networks leveraged for data management are often clustered, raising possible risks of single point collapse in the event of natural disasters. Blockchain is a decentralized, evolving, and revolutionary platform that has the ability to radically revolutionize, reshape and change the way of data accessibility in the healthcare sector. This chapter explores how the use of blockchain for data processing systems and accessibility in healthcare will stimulate advances and deliver substantial changes. This chapter presents a case study to show the practicality of blockchain technology for various healthcare applications.

Keywords Blockchain technology · Data · Management · Accessibility · Healthcare

O. Ali (✉) · A. Jaradat · S. Rotabi
Department of MIS, Business, and Administration College, American University of the Middle East (AUM), 54200 Egaila, Kuwait
e-mail: Omar.Ali@aum.edu.kw

A. Jaradat
e-mail: Ashraf.Jaradat@aum.edu.kw

M. Ally
School of Management and Enterprise, Faculty of Business, Education, Law, and Art, University of Southern Queensland, Springfield, Australia
e-mail: Mustafa.Ally@usq.edu.au

S. S. Muthu (ed.), *Blockchain Technologies for Sustainability*,
Environmental Footprints and Eco-design of Products and Processes,
https://doi.org/10.1007/978-981-16-6301-7_5

1 Introduction

Healthcare industries have been revolutionized due to the rapid developments in technologies, because of which there have been significant improvements in e-health/medical records (EHR/EMR), insurance information, and prescription drug data [35, 75]. The sophisticated medical devices can be used to gather critical patient data, offer information on disease symptoms and trends, create automation in workflows, assist in remote caring and put patients in greater control of their treatments and their lives [2, 8, 9, 91]. It is possible to monitor patients in real-time using advanced medical devices. In addition, they decrease the need to pay visits to hospitals for routine health check-ups. Connected home health monitoring systems are now available that help in decreasing hospital stays of patients and their readmission expenses. Diagnosis is facilitated by the medical devices embedded with sophisticated technologies by means of alerts and trigger notifications before the disease reaches a serious stage [7, 15, 50, 108]. Data can be collected through the sensors incorporated in the different parts of a patient's medical apparatus, which is then sent to the hospital where it is analyzed by a health professional for potential irregularities.

There is no doubt that technological advancements have bought continuous innovations in the healthcare sector [6, 48]. However, it has become very difficult to safely manage EHR/EMR because the data is transferred to different medical facilities [81]. There is centralization of the majority of the healthcare systems, which are at risk of single point of failures and information breaches because of the increase of cybersecurity attacks [80]. There are serious repercussions of the leakage of patients' personal and vital information. In addition, the existing medical systems do not offer transparency, immutability, audit, trustworthy traceability, privacy, and security while handling EHR/EMR [99]. Taking into account these issues in the existing healthcare systems, blockchain technology is a latest development that can be used to solve these issues [22, 30, 34, 60]. Blockchain adoption can give rise to annual savings of approximately \$100–\$150 billion by 2025 in expenses linked to data breaches and by bringing about a decrease in frauds and fake products [103].

Healthcare data management functions can be facilitated by the emerging blockchain technology as it offers unparalleled data efficiency and enforces trust [24, 25, 28, 47, 66, 87]. There are various key built-in features offered by the blockchain technology, for example, transparency, authentication, decentralized storage, data access flexibility, security, and interconnection, and this is why the technology is used extensively for healthcare data management [39, 96].

Smart contracts are employed by blockchain technology, which comprises of terms and conditions that are accepted by all the healthcare partners that are part of the network; thus, eliminating the need for an intermediary [7, 38, 105]. This decreases unnecessary administrative expenses. There are essentially three concepts on which blockchain is based, which are peer-to-peer networks, consensus methods, and public key cryptography [79]. There are three categories of blockchain, depending on their permissions: public, private, and consortium blockchains [9]. Any person connected to the Internet can take part in the consensus process in the public blockchains. Incentives and encrypted digit verification are included in the public blockchains through proof-of-work or proof-of-stake methods. The public blockchain system is fully transparent, where the identity of every participating individual stays pseudo-anonymous. Only a single organization can control the network in a private blockchain, which is why a trustworthy agent is required in this kind of blockchain to obtain consensus. The benefits of both public and private blockchain networks are integrated in the consortium blockchain. This kind of network is only appropriate for those organizations that seek to achieve efficient communication between each other. Depending on the specific requirements or use case scenarios, any kind of blockchain network can be used by healthcare organizations because of the benefits and drawbacks inherent in each of them.

In this chapter, the main features of blockchain technology are discussed, in addition to the key benefits they offer to healthcare organizations. In addition, the main opportunities provided by this technology to the healthcare industry are analyzed. A particular health case study is also discussed to demonstrate the applicability of blockchain-based healthcare systems.

2 Blockchain Overview

Blockchain technology is a decentralized digital ledger that provides an opportunity to record and share information in a community [8]. Each entry is transparent and searchable, thereby enabling community members to view its history. The cryptology in blockchain substitutes third-party intermediaries as trust keepers, while all participants run complex algorithms to certify the integrity of an entry. This technology can provide a new model for HIE by attempting to decentralize EHRs, thereby improving system efficiency and security [86]. Although blockchain technology is not a panacea, this technology has been led to a rapidly evolving field in the industry. Blockchain is important because it brings trust to P2P networks. The key component of blockchain technology includes consensus mechanism, distributed ledger and public key cryptography [33]. These components communicate and coordinate over a distributed network of devices owned and maintained by multiple entities.

The blockchain platform adopts a decentralized architecture, in which all network members achieve the required application purpose. A system state perceived in one machine is replicated through the execution of consensus mechanism logic and P2P networking protocol to all other devices in the network [33]. The replicated state information is stored in the context of blockchain, which is referred to as distributed ledger and is uniformly managed by the members of the network. Thereafter, the public key is used with a hash function to create a public address that users use to send and receive valuable assets [3]. The private key, which is used to sign a digital transaction to ensure that the transaction's origin is valid, is maintained confidential [3]. Each block in the blockchain consists of at least one transaction, signature of the block validators, and reference to the previous block along with block headers.

Blockchain provides opportunities for the standard architecture integration to radically transform our method of addressing various disciplinary systems issues, such as the Internet of Things (IoT) [67, 76], supply chain management [32] and Industry 4.0 [31]. This advantage is due to the decentralized nature of blockchain technology, in which many users own an entire database of a particular system. These decentralized database systems based on blockchain can reduce one of the cheating sources of database manipulation. Blockchain technology and cryptocurrencies have received significant industrial and academic attention [106]. Notable factors and opportunities will revolutionize the healthcare sector through its integration with blockchain technology [72].

3 Key Blockchain Technology Features

Blockchain technology comprises several features that can be utilized by the healthcare industry. These features are intrinsic to the healthcare systems, and they can be applied to a broad of systems and industries. The features to be discussed specifically in this section are Decentralization, Transparency, Immutability, Traceàbility, Trustless, Persistency, Anonymity, Auditability, security, Authentication, and Assurance.

3.1 Decentralization

Decentralization represents a very predominant feature of blockchain technology [7]. Due to this inherited feature, blockchain enables the sharing of database directly in a distributed ledger without intermediaries, where transactions are processed and stored by the network nodes [46]. As a transaction in the blockchain network can be conducted between any two peers without the need for central agency authentication, trust concern and server cost will be reduced through the usage of various consensus procedures [63]. Moreover, the decentralization nature of blockchain architecture exhibits decentralized storage and data management, which can be the basis for the enhanced security and authentication of the information stored within healthcare systems [60]. Decentralized storage and data management is achieved by spreading the storage from one major server into multiple servers through blockchain's ledger [16]. This will offer faster access to medical data and improve their quality as well [60]. Blockchain technology can be utilized as well by Mobile healthcare applications' developers to offer a secure and private sharing of patient data in a decentralized manner [84].

3.2 Trust and Transparency

Trust is one of the major concerns that exist in any traditional centralized healthcare system, where patients need to put their confidence in this system [63]. Blockchain technology through its decentralization nature brings trust to P2P networks by offering such components as a consensus mechanism, distributed ledger, and key cryptography [43, 46]. These components will eliminate the need for a trusted third-party intermediary and reduce the risk of failure.

Blockchain technology facilitates transactions processes to occur between unknown parties that do not trust each other. Although, the distribution of ledger across multiple nodes in blockchain network and the update of this ledger through consensus protocols will ensure the validity of transactions in an untrusted environment [46]. The utilization of consensus protocol as Proof-of-Work (PoW) and Practical Byzantine Fault Tolerance (PBFT) helps the network participants to communicate with each other to reach into an agreement by assuming that all honest nodes will have the same exact copy of the ledger [46]. This node clustering approach where each cluster can have a trusted manager that maintains the ledger can have a significant impact on providing a secure network for healthcare transactions [21, 46].

Accordingly, a new decentralized model of healthcare information sharing and exchange can be obtained to replace the traditional centralized healthcare system, which improves the system efficiency and information security level [43]. In fact, blockchain could provide a trustworthy infrastructure, which allows data creators and consumers to authenticate and preserve their data [49]. For instance, if the blockchain consists of several blocks, then the last one will contain a cryptographic hash of the previous block. Moreover, the information of the current block will be used to create a new block. Therefore, this chain of connectivity among the participated blocks will detect and eliminate any false manipulation of the transaction information and assure the digital form of verification and ledger approval [46, 49].

Transparency is another significant feature that can be offered by blockchain technology toward healthcare applications industry. Data replication is a mechanism in such distributed system that allows decentralized nodes to access a shared replicated data ledger [46]. In this mechanism, the transaction data in the blockchain network is replicated in the nodes and recoded as a chain where all the transactions linked together to go all the way to the first transaction. Accordingly, high transparent and secure transactions will be observed as any transaction changes that might occur on the network will be publicly visible [46]. Due to the transparency nature, blockchain technology has the capability to establish a secure and robust transparent framework for storing medical records. This framework offers quality services for the patients, and it reduces the treatment cost as well [43]. Moreover, healthcare applications can utilize the transparency and the tamper proof features of blockchains to handle public transaction logs, which include timestamp information and can be publicly disclosed and checked by all the participants in the blockchain network [43]. Therefore, through blockchain transparent and open nature, a trusted atmosphere, and a general acceptance about the usage of distributed healthcare applications by the healthcare providers will be achieved [4].

3.3 Traceability and Auditability

The blockchain architecture conforms to the auditing and tracing features [7, 43, 46]. In fact, the distributed and transparent nature of blockchain technology makes it easier to audit and trace back complex transactions to its origin [46]. Any transaction that might occur in a blockchain network is recorded and validated by a digital distributed ledger and timestamp. Auditing and tracing previous records are possible by accessing any node in the network [63]. Appropriately, blockchain technology can be utilized in healthcare and pharmaceutical industries to trace drugs and patient data to overcome such a big problem as drug counterfeiting that can put a patient's life in danger [63].

According to research by [17], auditability refers to all transactions in a blockchain that are maintained in chronological sequence, including the previous block's hash and storage of the hash, which is intended to connect the next block when it is added. Transactions can be readily confirmed and monitored using this technique. In addition [63], define auditability as the transactions that take place in a blockchain network and are recorded in a digital distributed ledger and authenticated by a digital timestamp. As a consequence, every node in the network may be accessed to audit and track past data [98]. In Bitcoin, for example, all transactions can be tracked iteratively, facilitating auditability and transparency of the blockchain's data state. However, if money is flowing through many accounts, it becomes extremely difficult to track it back to its source. In sum, blockchain technology will lead to ensure that the overall credibility of healthcare applications. Also, it refers to maintain a log of all the transactions [88].

3.4 Persistency and Immutability

Immutability is the blockchain feature that makes the modification of transaction records once stored in a particular block in the chain is impossible [7, 46, 63]. The data replication mechanism and distributed storage method allow a transaction to be stored in blocks through the entire blockchain network. This creates several redundant sources to verify the authenticity of the original transactions as each block in the chain is linked to the previous one using a cryptographic hash function. By having this redundancy, any attempt to modify a recorded transaction in a particular block will affect the subsequent blocks in the chain. Subsequently, a malicious actor requires to computationally change all the previous blocks in the chain to modify a transaction at a particular block, which is considered astronomically difficult considering the amount of work that needs to be done. Therefore, immutability, assurance, and security are obtained where any manipulation or forgery of data will be detected by the network [43, 46, 63]. In fact, blockchain provides the infrastructure by which truth can be measured and it enables participants to prove their information are authentic and unchangeable. As a result, blockchain is recognized as an immutable distributed ledger [63].

3.5 *Privacy and Anonymity*

Blockchain is recognized as a technology that aims to provide safety and privacy to sensitive personal data due to the facility of allowing users to do transactions with generated addresses instead of using a real identity [63]. Due to the decentralization and trustless environment of blockchain, no central authority is monitoring and recording users' private information. The interaction with the blockchain network is based on random address generation, and hence, the user can have many addresses within a blockchain network to avoid the reveal of his identity. This feature and through the immutable ledger technology could provide a significant impact on obtaining confidentiality and integrity of patient records while integrating and sharing them among different healthcare facilities [63]. However, some recent studies on the Bitcoin platform have indicated that the history of a transaction can be linked to reveal a participant's true identity [85]. This information leakage vulnerability is observed because all the public keys details are visible to everyone in the network, and hence, privacy prerequisites should be identified at the initial stage of blockchain applications [63].

4 Benefits of Blockchain in Healthcare

With the advancement of internet technology usage, healthcare industry is currently facing several unique requirements that need to be handled by information technology and medical communities to assure quality services for the beneficiaries [60]. These requirements are associated with security, privacy, interoperability, data sharing, and mobility data access [60]. Due to dissemination need of patient's medical data and the existence of centralized data storage, traditional healthcare applications and data management approaches fails to provide a proper solution to handle these requirements [60]. Accordingly, the adaptation of such technology as blockchain, which provides decentralization transactions processes with secure and trustworthy information access and sharing paradigm, into healthcare application industry has become a necessity. This necessity is outlined toward the advancement of digital healthcare industry by overcoming the limitation of the traditional healthcare applications [46, 60]. Issues existed in the traditional client-server healthcare architecture as patients' records vulnerability, real-time patient data access, lack of drugs traceability, healthcare data fragmentation, access control, patient data privacy, and healthcare applications security can be solved by the integration of blockchain technology into healthcare applications and through the usage of decentralized, shared, immutable, and transparent ledgers [46].

In recent years, there is a considerable optimistic belief that blockchain will revolutionize the healthcare industry [53]. Blockchain technology applied to healthcare sector can offer new and effective improvements opportunities. It also can play a crucial role in several healthcare applications like public healthcare data management, electronic health records, biomedical, laboratories, pharmaceutical, automated healthcare services logistics, online patient access, sharing patients' medical data, drug counterfeiting, clinical trial, precision medicine, smart health, and real-time health monitoring, and mobile health [7, 21, 40, 46, 53, 60].

The major benefits to be discussed specifically in this section correspond to blockchain features that handle the healthcare unique requirements that comes in the form of having a decentralized structure, allowing interoperability, security, authentication, and integrity. These benefits are discussed next.

4.1 Decentralized Medical Data Management

The centralization storage of healthcare applications has become an obstacle to provide quality services, disease diagnoses, and advance care for patients [46, 60]. Current medical data management are based on client-server architecture where hospitals are the primary keepers of the data [46]. Centralized data management creates a challenge for patients to have a unified view of their medical history since patient data could be stored at different hospitals and clinics. It also creates a challenge for healthcare experts to make precise disease diagnoses due to the lack of having a comprehensive informant about the patients' medical history. Thus, blockchain can support the advancement of healthcare applications such that decentralized data management can be the backbone of this advancement [4]. Patients and health professionals can have controlled access to the same medical records without giving anyone the rule of central authority over the global health data [4, 49]. Moreover, this decentralization will also motivate health service providers toward medical records standardization that will offer easy collaboration and follow-up between health's professionals and service providers [49]. As result, an overall improvement of healthcare services would be obtained.

4.2 Data Protection

Storing patients' medical records is very essential in healthcare applications and services. These records are often very sensitive and usually are a main target for cyberattacks. Therefore, such records should be kept secure and private [51]. Due to the decentralization and immutability features of blockchain, no single data point can be hacked to steal immutable patient records [51, 63]. Thus, once the medical records are stored at blocks network, no one can corrupt, change, or retrieve these records [4, 53, 63]. All the medical data on the blockchain are encrypted, timestamped, and tamper proof [4, 53]. Furthermore, some cryptographic schemes and algorithms, such as attribute-based encryption, and identity-based encryption and signature, are utilized within blockchain architecture to strengthen the protection of patients' identity and records [4, 89]. Therefore, blockchain technology is capable of establishing a robust and secure framework for protecting patient's medical history and identity. Consequently, patients and healthcare service providers will have higher confidence in using such blockchain-based healthcare data management systems, which eventually lead to an increase in the level of medical data sharing and hence improve the overall healthcare services [51].

4.3 Ownership of Medical Data

Patients and healthcare providers want to have control of their own data and how the data will be used by others. Therefore, the adaption of blockchain technology in healthcare applications and due to blockchain's persistency and immutability feature, authenticity, and ownership of patient health records would be ensured. No one can steal the stored patient records in the distributed ledger without the consent of the owner and that it would be impossible for that information to have tampered [4, 46, 63]. Blockchain assures data ownership through the usage of unique cryptographic layers and protocols as well as through the utilization of well-defined smart contracts [4, 60, 63].

4.4 Control Pharmaceutical Supply Chain

Blockchain technology due to the immutability, traceability, and transparency features can support patients and healthcare professionals to overcome the drug counterfeiting problem [39, 51, 63]. Blockchain-based pharmaceutical supply chain application will be able to trace the pharmaceutical row materials offers by suppliers to the finished products recommended to the patients in an immutable and shared distributed ledger [51, 63]. In addition, blockchain can be a solution to handle the annual loss for pharmaceutical organizations that occurs due to the drug counterfeiting problem [63]. This solution is achieved since all the transactions added to the distributed ledger are immutable and digitally timestamped and that makes it possible to trace a product and make its records tamper proof. Moreover, by integrating blockchain-based pharmaceutical supply chain applications with IoT anti-counterfeit devices, any counterfeit drug in the supply chain network will be detected throughout participants' data verification [51]. According to this technological integration, a new open standard technology that migrates the pharmaceutical supply chain application developments toward transparent, shared, and trustworthy architecture, which as a result will improve the quality of global healthcare services and collaboration.

4.5 Availability and Accessibility of Medical Data

Due to the replication feature of blockchain, where data from different sources can be replicated in different nodes of the network, the availability of medical data that are stored on the blockchain is robust and resistant against data loss or data corruption [4, 49, 51]. This would motivate patients and healthcare providers to securely share access to their selective medical data with any trusted third party [49, 51]. Accordingly, blockchain through the usage of public and private keys can provide an easy and private data access platform for patients and healthcare professionals to overcome the overall regulations, privacy concerns, and legal difficulties that we awfully face in the traditional healthcare data sharing and aggregating the environment [49].

Blockchain technology would ensure the continuous availability and near real-time access to medical data [39]. In specific, the decentralization feature of blockchain technology will improve the performance of healthcare near real-time data transactions and processes [39, 46, 51]. For instance, blockchain technology by providing near real-time claim processing, where preauthorization and eligibility verification processes are encrypted and shared in the blocks, would improve the efficiency of the health insurance claim process. In such cases like emergency, remote patient monitoring, and medical intervention, blockchain technology would provide an active data sharing and feedback loop that allows automatic delivery notifications between all the involved medical professional parties [51]. As result, availability and convenient access in real-time to the medical data would enhance the professional, and educational coordination and collaboration among in healthcare sector, which ultimately would the quality of services and treatments offered to the patients.

5 Blockchain Application in Healthcare Sector

Healthcare sector represents a problem-driven, data interoperability, and personal intensive domain, such that distributed data access, data sharing, trust, and data ownership are critical aspects within the sector processes and operations [39]. Therefore, wider applicability of blockchain technology could overlay its ways into different healthcare aspects. In fact, the features of blockchain technology may enhance and offer unique solutions to several healthcare applications. This section discusses the applicability of blockchain technology toward enhancing the services and operations offered in varied healthcare applications such as medical data management, mobile health, pharmaceutical tracing and supply chain, and health data analytics and research.

5.1 Mobile Health (mHealth)

mHealth that represents mobile applications and remote monitoring facilities is one of the growing fields in healthcare applications [60]. It involves the collection of biomedical data through the usage of such devices as miniaturized sensors, low-power body-area wireless networks, and pervasive smartphones to remotely monitor the status of the patient outside traditional healthcare environment [4, 22, 55]. Similar to the broad healthcare centralized server systems shortcomings, mHealth suffers as well from such shortcomings as access control, authentication, user trust, and data sharing [78]. Accordingly, incorporating blockchain technology into mHelath applications can resolve such shortcomings and improve security and quality of machinery [22, 60]. Several m-Health applications have been developed based on blockchain incorporation. For example, a smartphone application for cognitive behavioral therapy was developed by a group of researchers [44]. This application provided a secure and tamper-resistance network, where patients can monitor their own care by recording and remotely sending their medical data to the healthcare providers. Healthcare Data Gateway (HGD) is another smartphone application that has been developed for organization of patient data [99]. Moreover, remote healthcare systems, such as the ones presents in [35, 71], were proposed on the basis of smart contract in the Ethereum blockchain model to enable real-time patient monitoring and to provide a secure operation process for the monitoring devices [22, 43].

5.2 Healthcare Data Management

The management of medical and healthcare data as it includes data storage, access control, and data sharing is an important aspect of the healthcare systems industry [4, 43, 52]. Managing medical data in a proper fashion would improve the healthcare outcomes by offering efficient communication and collaboration among decentralized healthcare providers and specialists that allows the sharing of holistic views of patients' medical records and personalized treatments [4, 52]. Proper data management is also a very important aspect for healthcare industry to achieve efficient and cost-effect operation processes [52]. Due to the confidentiality and subsequent trust issues, managing and dissemination healthcare and patient records are challenging tasks that raise several security and privacy concerns [4]. Therefore, encapsulating blockchain technology with healthcare data management approaches will enable the efficient storage and sharing of medical data while ensuring data integrity and patient privacy [52]. Accordingly, several researchers were focusing on developing blockchain-based Electronic Health Record (EHR) systems to facilitate patient-centric data storage and data sharing among varied healthcare providers [4, 43, 52]. For example, Guard-time Company utilizes a blockchain-based platform to secure the records of more than one million patients in Estonia [4]. The Gem Health Network (GHN) and HealthChain are other examples of EHR system that allows different healthcare practitioner to share access of the same data [4]. HealthChain HER application was developed as a permissioned blockchain network that incorporates smart contracts to control access privileges and authorization processes on the network [5].

5.3 Pharmaceutical Tracing and Supply Chain

Medical supply chain represents one of the most important processes in the medical industry [4]. It is the process that starts from raw materials selection and drugs productions to the different stages of storage and distribution, efficient monitoring and tracking, and proper detection of counterfeit drugs to ensure optimal usage of drugs supplies [18, 51, 52]. One of the growing concerns for this healthcare field is that the delivery of counterfeit or substandard medications [51, 52]. Counterfeit medications could consist of incorrect dosage of active ingredients and/ or inactive ingredients that may cause harmful allergic reactions to the patients [51]. To address this concern, it becomes essential that drug supply chain provides a robust and consistent tracing mechanism across industries. Blockchain technology has been identified as a perfect fitted tracing solution, where drugs related information is obtained in a safe and tamper proof open system that allows multiple healthcare parties to access this information [51, 52]. Accordingly, many blockchain-based medical supply chain applications, such as MediLedger [52], Counterfiet Medicine project [14] and Ambrosus [52], have been proposed to track the pharmaceutical supplies. The main idea of these applications is to record every transaction related to the drugs prescription on the blockchain network to which all the participated parties are connected and had shared access to the stored records [4].

5.4 Health Insurance Claims

Financial aspects of medical treatments are very important in the healthcare industry. Unfortunately, this financial aspect in healthcare industry exists with errors, fraud, and inefficiency in terms of trust and transparency, which can be very costly to the healthcare industry [51, 52]. Blockchain technology through its decentralization, immutability, and auditability features offers an optimized mechanism of direct and transparent links between patients and insurance claim companies. For instance, unlike the traditional healthcare database systems that support create, read, update, and delete functions, the immutable auditing path of blockchain only supports the create and read functions [51]. Thus, the robust and unchangeable ledger of blockchain is more appropriate to record critical information as insurance claims [52]. In addition, blockchain's provenance feature allows the traceability and verifications of the insurance claims' origins, hence, insurance's transaction data reusability will increase [51]. Accordingly, several research papers identified that blockchain-based insurance claims applications are a very promising area for further research and industrial developments [4, 39, 51, 52]. However, very limited examples of prototype development for such system are actually implemented in this area [4, 51]. Among them, the MIStore application [107] that represents a good example for providing an actual blockchain-based application to support insurance claims storge throughout the deployment of Ethereum blockchain platform [4].

5.5 Health Education and Clinical Trials Research

The quality, sensitivity, and credibility concerns in healthcare industry toward providing best practices and patient's treatments emerge the need for blockchain technology adoption in such areas as healthcare researches, educations, and medical trials [4, 18, 39, 51, 52]. In fact, as the healthcare industry aims to maintain and support patient's health, it does not accept the lack of judgment or inaccurate medical information and practices [52]. Blockchain technology throughout its traceability, digital credentialism, and immutability natures can have a huge positive impact on the outcomes from healthcare researches, educations, and medical trials. By incorporating blockchain technology into these healthcare areas higher confidence of the quality and credibility of the acquired healthcare information and attributes, and of the returned outcomes, which will improve in general the quality of treatments and practices offered to the patients [4, 18, 39, 51, 52]. This can be achieved by maintaining digital credentials and traceability of the healthcare educational and clinical trials contents and outcomes [18].

Even though none of the research works in this healthcare filed provided an actual blockchain-based prototype implementation and validation in practice, the potential utilization, and incorporation of blockchain was identified to address the traceability, ownership, and management of clinical trials, educational and medical research contents, and physician's digital credentials [39, 51]. For instance, a study done by [54] addressed the scope of utilizing blockchain for medical education in Russia, while [58] study discussed how the utilization of blockchain technology can increase the credibility clinical trials data, and hence, the obtained results as well [51]. Moreover, there are other potential healthcare applications areas such as dental industry and meaningful use can benefit from the utilization of blockchain technology, where further intensive research efforts and actual prototype development is required in practice [4].

6 Case Study (e-Health Estonia)

Estonia became the first country to introduce blockchain on a national level. The country of under 2 million population has been at the forefront of innovation in digital society for the last 20 years and is the only country where a majority of citizens carry a PKI smart card providing access to over 1000 electronic government services that are actively used. Electronic patient records are a critical component of these services and by integrating blockchain technology it becomes possible to provide an independent forensic-quality audit trail for the lifecycle of those patient records, making it impossible for anyone who gains access to those records to manipulate information and cover their tracks.

The Estonian e-Health Foundation has been operating since 2005. About 95% of health data is digitized in Estonia and over 300 M-health events are saved in Blockchain. Estonia is placed to lead preventative medicine, patient self-treatment, and industry efficiency [64]. Over this time (Fig. 1).

In 2016, the e-Health Foundation teamed up with Guard-time, a company that specializes in data security. Guard-time helped the foundation to introduce KSI (Keyless signature infrastructure), a blockchain technology that provides large-scale data authentication without relying on a centralized trusted authority, and a zero-trust system that can provide a formally verifiable mathematical proof of the correctness of operations. Guard-time was founded with a mission to "make the world's information universally reliable, without reliance on the risks of human trust". The company provides a blockchain-based system to Estonia to secure 1 million health records [64] (Fig. 2).

Now the project has over a million records of patients and their data. KSI infrastructure provides high security of medical data, its safety, and integrity. With every alteration of medical information, the KSI blockchain automatically creates an updated record, which prevents any uncontrolled data manipulation, whether it is an edit, addition, or deletion. Guard-time HSX APIs provide easy to use APIs for building distributed, secure, and compliant healthcare applications.

99%

prescriptions were issued digitally

95%

medical records were digitized

500 000

requests are sent by doctors annually

100%

invoices were issued digitally

Fig. 1 Estonian e-Health data [64]

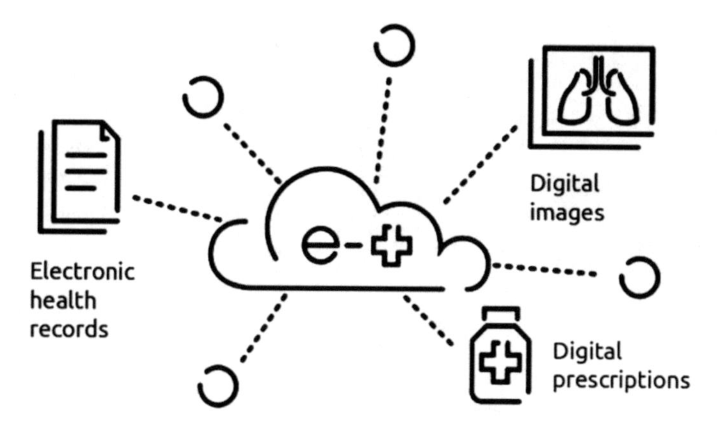

Fig. 2 Blockchain interaction with Estonia's e-health systems [64]

The system provides instant access to primary care information, personal care pathways, and medication adherence support through their smartphone. The platform is designed to deliver immutable proof of health data provenance and integrity, patient data rights management, and automated verification of medication adherence.

Estonian e-Health Foundation uses Oracle technology to process and store the patient records and Guard-time's KSI blockchain will be integrated at the Oracle database engine, providing increased security, transparency, auditability, and governance for electronic systems and lifecycle management of patient records. KSI instrumented records will be irrefutable [36].

Estonia, Hungary, and Iceland, together with AstraZeneca Estonia are participating in a pilot of Guard-time's Vaccine-Guard, its newly developed platform to support the global COVID-19 vaccination program, ensuring reliable vaccines, vaccination certificate interoperability, and pharmacovigilance. The product, built on KSI blockchain technology, is based on a six-month collaboration with the Estonian Government and World Health Organization.

The network is an open platform for public health authorities, hospitals, citizens, certificate providers, vaccine manufacturers, border guards, and others to securely and reliably share information across systems and borders.

The solution provides a feedback loop between all participants in the network for usage cases as diverse as counterfeit detection, vaccine allocation prioritization, and pharmacovigilance while employing leading privacy and security features to protect patient and other sensitive information Guard-time, Europe's leading deep tech company, has announced that Estonia, Hungary, and Iceland are the first countries to sign-up to pilot Vaccine-Guard [1] (Fig. 3).

AstraZeneca Estonia will participate in vaccine-Guard product testing enabling an end to end solution with a feedback loop between manufacturer, care provider, citizen, and public health authorities.

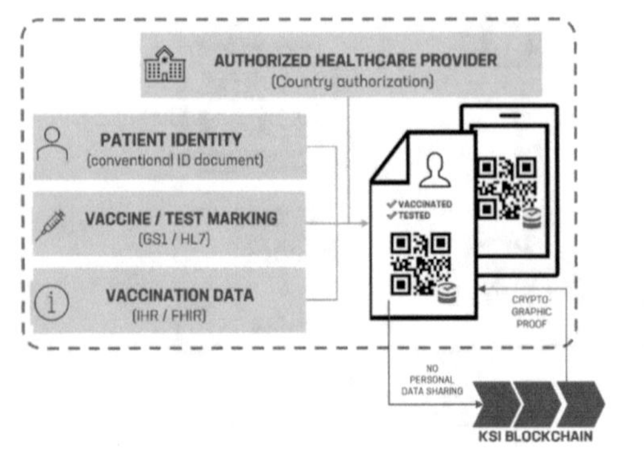

Fig. 3 Vaccine-guard adds immutable attestation to the certificate [1]

The product is based on a six-month collaboration with the Estonian Government and World Health Organization (WHO) with additional governments expected to join the pilot network in the near future [1] (Fig. 4).

Vaccine-Guard will be the first solution adopted by national health authorities that links decision-critical data like vaccinations and authentic vaccines across multiple systems integral to the successful delivery of the COVID-19 vaccination program and enabling global travel.

Vaccine-Guard will provide proof of critical data accuracy, from verifying the vaccine against authentic vaccine data repository, managing, and monitoring compliance with national and local mandates on distribution and administration to priority groups for inoculation, to patient verification and eligibility, and real-time updates to health authorities, giving them better insight into the vaccination program progress and success. It will deliver automated aggregated reports from vaccination sites,

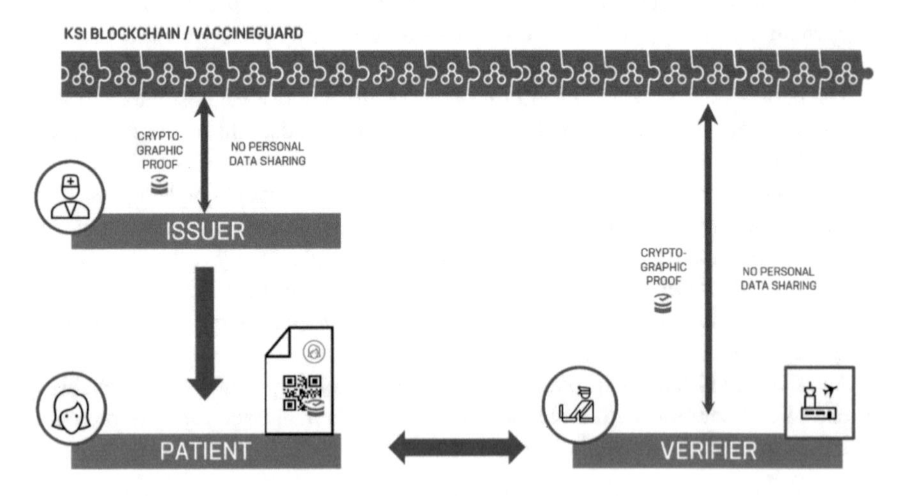

Fig. 4 Workflow of the certificate [1]

automated monitoring of stock and vaccinations, and provide the ability to facilitate adverse effect reporting, supporting investigations around this more quickly [36].

Guard-time is an integral part of the Estonian Government's recent offer to WHO and its member states to share its experience and that of its key companies in healthcare digitization that use distributed information architecture and interoperability, and how this could be used in the fight against COVID-19.

Jüri Ratas, the Prime Minister of Estonia said "*The pandemic has shown the world that in crisis we need to have a global anchor of trust like the WHO. The World Health Organization can play a critical positive role in global digital health governance. Our pilot project presents an extraordinary opportunity for the Estonian community to help the world in the fight against COVID-19*" [36].

While there are many aspects to any e-Health system, the Estonian Case Study clearly demonstrates how blockchain technology can be successfully incorporated as the underlying infrastructure on which essential systems such as access to patient records and billing systems can be built thereby exploiting blockchains inherent attributes of security, privacy, and trust. The blockchain then becomes the backbone for digital health, incorporating data from patient-based technologies to provide a pool from which authorized users, such as providers and patients, have access. In the case of the Vaccine-Guard program, all of the data could be stored in a decentralized manner, with no single entity storing or having singular authority to access them. Blockchain-enabled anti-tampering capabilities during manufacturing, and the supply and dispensation system could mitigate against any counterfeiting.

Each person in Estonia, which has been at a doctor, has their own online e-Health record that can be tracked. The National Health Information System integrates data from Estonia's different healthcare providers, creating a common record for each patient (since 2015, over 95% of data generated by hospitals and doctors has been digitized). This gives the doctors easy access to the patient's electronical records (i.e., test results, X-ray images). Patients have access to their own and their children's records. By logging into the Patient Portal (ID-card/m-ID), they can review their past doctor visits, current prescriptions, receive general health advice, etc. Blockchain technology is also used in the Estonian national health information system to ensure data integrity [62].

Patient data is securely accessed through secure digital authentication and signing. The infrastructure also allows forwarding data by using an encrypting key pair: a public encryption key and a private decryption key. In Estonia, this technology is used for electronic identity (ID card, mobile ID, and digital ID). To ensure the transparency and public accountability of the system in e-Health, the patient can control the access to the data by allowing or restricting access to it and also see everyone who has access to the data. Every update to healthcare records and every access to healthcare records is registered in the blockchain. That makes it impossible for the government or doctors or anyone to cover up any changes to healthcare records [93].

7 Processes Improved Through Blockchain

It was briefly explained in this section how blockchain technology can offer significant improvements for streamline healthcare data management functions.

7.1 Patient Record Management

Every patient is unique; hence, it is usually not possible to apply similar strategies because of inter-individual differences [19, 37, 70, 94]. Therefore, medical records should be completely accessible so that personalized care can be provided to patients [26, 29, 61]. The issue that arises with sharing medical records among the medical community is that most of the existing medical systems do not ensure security, privacy and trust [90]. In addition, patients are not able to claim complete ownership of their medical records as it is possible to alter or delete information from medical records. In case patients shift to another facility, the tests done earlier need to be repeated, because of which there is an increase in costs. These issues can be resolved through blockchain technology [13, 58, 94, 104]. Data is stored on a decentralized peer-to-peer network that can only be accessed through smart contracts. It is possible to transfer this data from one hospital to another without the risk of misuse. It can then be used by doctors to acquire information about the patient's history, which helps them understand their case better and offer them appropriate treatment. Blockchain also prevents further costs as patients do not have to repeat diagnostic tests that were carried out previously [30]. All copies of patient records are stored on different nodes that are part of the blockchain network, which ensures that they remain accurate and transparent [68].

7.2 Maintaining Consistent Permissions

Healthcare practitioners need to have quick access to patient data in medical emergencies. If there are inconsistent permissions, a patient's data access can face hurdles in a medical emergency, which puts the patient's life in danger [45]. Two solutions can be offered by blockchain technology with respect to seamless and secure permission management. There are predetermined rules in blockchain-based smart contracts that have been approved by all individuals that are part of the contract, which help in offering access. It is possible to customize these contracts so that the workflows can be automated [56, 57]. Patients can use cryptographic keys to regulate access control. Healthcare data can be unlocked by the master key that is available to every patient. Patients can also share a copy of their data with medical practitioners or hospitals whenever required. Smart contracts can also be used to add read and write access rights. Blockchain-based smart contracts and cryptographic keys help in decreasing the errors that occur because of human negligence. In addition, collection time for patient data can be decreased through blockchain.

7.3 Protecting Telehealth Systems

Though telehealth systems are capable of getting past geographical constraints in healthcare, they are at risk of experiencing cybersecurity attacks. When the virtual connection that is formed between a doctor and patient is affected, the sensitive information regarding the patient gets compromised, for example, routine data transmissions, household activities, internal brand information, etc. [97]. Telehealth systems' effectiveness is essentially dependent on the way the security and privacy issues are managed. Blockchain plays an important role in offering security, trust, and privacy in telehealth systems [39, 74]. It facilitates the development of seamless data exchange in the absence of an intermediary that enhances customer confidence in the telehealth systems. Doctors can use it to store extensive history of patient, laboratory outcomes, and records of treatment/procedures in a decentralized, accessible, immutable, and traceable manner [39]. The integration of blockchain technology into telehealthcare systems faces obstacles in that it may increase the cost for patients living in remote areas where there is a scarcity of resources.

7.4 Clinical Trials and Precision Medicine

The trustworthiness of data obtained during the clinical trial research can be enhanced through blockchain technology. It makes sure that there is data integrity, due to which the issue of presenting incorrect clinical trial data can be handled [66]. It increases transparency and helps in enhancing the precision of data analytics that are carried out on the data obtained from the clinical trials. The use of blockchain for clinical trials research helps mitigate various issues like tracing and auditing of the clinical supply chain, patient recruitment, decreasing the overall time taken for carrying out trials, and reinstating the integrity of trial data. Blockchain can be used in precision medicine to handle the genomic sequences that help in proactively managing all types of diseases and illnesses that occur due to a genetic disorder [83]. It was found that 10% of the chronic diseases experienced by the adult population are genetically inherited. To treat these diseases proactively, an understanding of the individuals' DNA profiles is required, and for this, genomic sequencing is needed. However, it involves difficulties like interoperability of DNA data and the readiness of organizations to share this data with others that prevent this from occurring. Individuals are able to own and manage their data through blockchain-based DNA data storage. This removes the need to have centralized databases that are handled by third parties, which are at high risk of being hacked. Since there is secure storage of an individual's data on blockchain networks, they are able to share it with others for medical research, drug development, and public health research.

7.5 *Optimizing Health Insurance Coverage*

At present, the majority of insurance companies depend on centralized systems and technologies for the storage and processing of their data [39]. The life cycle of a health insurance policy typically involves various third parties or middlemen. In addition, there is sharing of information between different stakeholders of the insurance industry, because of which the entire process becomes time consuming and lengthy [20]. There are clear inefficiencies in the existing medical insurance system. An exceptional degree of transparency is offered by blockchain technology as it keeps a record of all transactions carried out in a decentralized, traceable, secure, immutable manner that cannot be tampered with. Blockchain is capable of providing a solution for the interoperability problem. Smart contracts make it possible to obtain transactions, agreement records, and other information in an automated manner, which can bring improvements in the administrative processes. In addition, smart contracts also help in identifying fraudulent or exaggerated insurance claims. Another advantage of blockchain technology is increased transparency as it makes it possible for physicians to view the health coverage of patients [74]. It also simplifies the health insurance process and enhances the precision of provider directory through consensus protocols. Therefore, blockchain is a very significant mechanism for the health insurance industries.

7.6 *Medical Billing Systems*

The conventional means of patient billing systems were exposed to different kinds of frauds in the past [97]. In addition, it takes significantly more time and resources to obtain billing information from the existing billing process [35, 61, 107]. Billing inaccuracies occur unintentionally because of the complex coding that is part of the medical billing system, such as inaccurate filings or duplicate processes. The medical billing systems can be optimized by integrating computer-assisted coding techniques with blockchain systems [39]. Blockchain is an emerging technology that has simplified the payment process and made it more secure in comparison to conventional billing methods, which most of the time makes significant delays in claiming the bills. The traditional payment systems caused even higher delays in bill payments of insurance claims. Such restrictions can be removed by blockchain which stores data in an immutable way. This allows insurance providers to make payments of insurance claims in a more rapid manner while reducing the need for additional time and resources.

7.7 Enhancing Privacy of Patients' Data

Stored information and patient-generated health data is very important [41]. Wearable devices have emerged due to technology innovations in the healthcare industry, for example, fitness bands, trackers, smart watches, and built-in body chips for monitoring patient health. Through these wearable devices, a greater amount of data is generated by patients. Solutions are offered by blockchain technology for the challenges that emerge with the increasing availability of healthcare data [99]. The digital health start-up, Health Bank that offers blockchain solutions by offering the following capabilities and facilities to the users [30]:

- Sharing and managing healthcare data and patient records;
- Obtaining private information about patients, such as their blood pressure, heart rate, health history, sleep patterns, medications taken, eating and lifestyle habits;
- Data storage and accessibility for medical research; and
- Storing and handling the data of facilities in a safe region.

Despite all these capabilities, users are still offered sovereignty of their information by the Health bank; data can be saved by the users, which can then be made available to medical researchers [30]. A valuable part is played by individual medical data stored at Health bank for patients and donors. Furthermore, blockchain makes it possible for researchers to obtain medical data for using it in their research, which could turn out to be monetarily beneficial for the patient and a source of side income if they agree to share their data for research [41]. A major contribution has been made by the Health bank to the health sector and the field of medical research, which has helped in the digitization of digital businesses and health programs (Table 1).

8 Conclusion

This chapter offered insightful discussions on the integration of the blockchain with healthcare systems. This chapter discussed how adopting blockchain technology in healthcare sector can lead to managing health data. Also, this chapter discussed the features of blockchain technology to show how it can unlock its full potential for healthcare data management. In addition, this chapter discussed the key opportunities offered by blockchain in the healthcare sector. Furthermore, this chapter presented Estonia e-Health as a case study to give an overview of how healthcare systems have been facilitated and complemented by blockchain technology. This chapter concludes that blockchain has the potential to reshape and transform the healthcare sector by bringing significant improvements in terms of operational efficiency, data security, healthcare staff management, and costs.

Table 1 Presents different health information systems along with the main challenges that they faced and illustrates how blockchain technology improved the processes within these systems

Health Information system	Process improved	Main challenge	Source
E-health records	• Shared decision making • Health data recording, storing, and sharing • Sharing of healthcare information for clinical and research purposes • Recording and sharing of contracts/agreements • Sharing healthcare data for administrative or economic purposes • Patients' collection, archiving, and sharing of healthcare data for clinical purposes • Retrieving information in the HER • Patient-controlled sharing of health data between healthcare providers • Sharing healthcare data between health institutions	• Interoperability • Access control • Access data • Data integrity • Identity management • Data provenance • Data privacy	Zhang et al. [100, 101, 103, 104], BlocHIE [82], Peterson et al. [70], Guo et al. [37], Xia et al. [94, 95], MedRec [16], Hussein et al. [42], Fan et al. [29], Mikula and Jacobsen [61], Dagher et al. [26], Anastasia et al. [13], Li et al. [58], Zhang and Poslad [104], Xia et al. [94, 95], Rahmadika and Rhee [73], Zhang et al. [100, 101, 103, 104], Farouk et al. [30]
Knowledge infrastructures	• Aid decision making by presenting knowledge	• Data integrity • Repudiation	Kleinaki et al. [39], Hasselgren et al. [53]

(continued)

Table 1 (continued)

Health Information system	Process improved	Main challenge	Source
Patient health records	• M-health data recording • Sharing healthcare data between health institutions • Automatic collection, storage, and patient-controlled sharing of personal health data • Collecting and sharing sensor data for clinical purposes • Managing access to personal health data	• Data integrity • Data provenance • Interoperability • Access control • Data privacy	Ichikawa et al. [44], Roehrs et al. [77], Liang et al. [56, 57], Zhang and Lin [100], Dias et al. [27], Liang et al. [56, 57], Uddin et al. [92], Farouk et al. [30]
Automated diagnostic service for patients	• Collection and storage of data about symptoms of dyslexia for the purpose of automated diagnostics • Decision-support and research	• Data integrity • Interoperability • Access control	Hasselgren et al. [39], Rahman et al. [74]
Administrative systems	• Sharing healthcare data for administrative purposes • Sharing healthcare data for economic purposes • Collection and storage of sensor data for remote patient monitoring purposes	• Data integrity • Data provenance • Identity management • Access control	Zhou et al. [100, 101, 103, 104], Mikula and Jacobsen [61], Griggs et al. [35], Hasselgren et al. [39], Farouk et al. [30]
Research support systems	• Establishing a patient-controlled marketplace for selling and buying of healthcare information for research purposes • Sharing healthcare information for research purposes	• Interoperability • Access control • Data integrity • Data provenance	Hasselgren et al. [39], Mamoshina et al. [59], Nugent et al. [65]

References

1. Aaviksoo A, Day G (2021) Vaccine guard. https://m.guardtime.com/files/Guardtime_VaccineGuard_Whitepaper_v2.pdf
2. Aazam M, Zeadally S, Harras KA (2020) Health fog for smart healthcare. IEEE Consum Electron Mag 9(2):96–102
3. Abramova S, Böhme R (2016) Perceived benefit and risk as multidimensional determinants of bitcoin use: a quantitative exploratory study. In: The 37th international conference on information systems, pp 1–20
4. Agbo CC, Mahmoud QH, Eklund JM (2019) Blockchain technology in healthcare: a systematic review. Healthcare 7(2):56 (Multidisciplinary Digital Publishing Institute)
5. Ahram T, Sargolzaei A, Sargolzaei S, Daniels J, Amaba B (2017) Blockchain technology innovations. In: Proceedings of the 2017 IEEE technology & engineering management conference (TEMSCON), Santa Clara, CA, USA, pp 137–141
6. Alam MGR, Abedin SF, Moon SI, Talukder A, Hong CS (2019) Healthcare IoT-based affective state mining using a deep convolutional neural network. IEEE Access 7:75189–75202
9. Ali MS, Vecchio M, Pincheira M, Dolui K, Antonelli F, Rehmani MH (2018) Applications of blockchains in the internet of things: a comprehensive survey. IEEE Commun Surv Tutor 21(2):1676–1717
8. Ali F, El-Sappagh S, Islam SR, Kwak D, Ali A, Imran M, Kwak K-S (2020) A smart healthcare monitoring system for heart disease prediction based on ensemble deep learning and feature fusion. Inf Fusion 63:208–222
7. Ali F, El-Sappagh S, Islam SR, Ali A, Attique M, Imran M, Kwak KS (2021) An intelligent healthcare monitoring framework using wearable sensors and social networking data. Futur Gener Comput Syst 114:23–43
10. Ali O, Jaradat A, Kulakli A, Abuhalimeh A (2021) A comparative study: blockchain technology utilization benefits, challenges and functionalities. IEEE Access 9:12730–12749
11. Ali O, Ally M, Clutterbuck P, Dwivedi Y (2020) The state of play of blockchain technology in the financial services sector: a systematic literature review. Int J Inf Manag 54:1–19
12. Al Omar A, Bhuiyan MZA, Basu A, Kiyomoto S, Rahman MS (2019) Privacy-friendly platform for healthcare data in cloud based on blockchain environment. Futur Gener Comput Syst 95:511–521
13. Anastasia Theodouli SA, Moschou K, Votis K, Tzovaras D (2018) On the design of a blockchain-based system to facilitate healthcare data sharing
14. Androulaki E, Barger A, Bortnikov V, Cachin C, Christidis K, De Caro A, Yellick J (2018) Hyperledger fabric: a distributed operating system for permissioned blockchains. In: Proceedings of the thirteenth EuroSys conference, pp 1–15
15. Awais M, Raza M, Ali K, Ali Z, Irfan M, Chughtai O, Khan I, Kim S, Ur Rehman M (2019) An internet of things based bed-egress alerting paradigm using wearable sensors in elderly care environment. Sensors 19(11):2498
16. Azaria A, Ekblaw A, Vieira T, Lippman A (2016) MedRec: using blockchain for medical data access and permission management. In: The 2nd international conference on open and big data (OBD). IEEE Computer Society, Los Alamitos, CA, USA, pp 22–24
17. Batubara FR, Ubacht J, Janssen M (2018) Challenges of blockchain technology adoption for e-government: a systematic literature review. In: The 19th annual international conference on digital government research, New York, pp 1–9
18. Bell L, Buchanan WJ, Cameron J, Lo O (2018) Applications of blockchain within healthcare. Blockchain Healthc Today 1(8)
19. Blockchain & healthcare—Drug traceability & data management. https://juraprotocol.medium.com/blockchain-healthcare-drug-traceability-data-management-traceability-259dd7c79c24. Accessed 10 May 2020
20. Breteau J (2020) The future of blockchain in health insurance. https://www.the-digital-insurer.com/future-blockchain-health-insurance/

21. Casino F, Dasaklis TK, Patsakis C (2019) A systematic literature review of blockchain-based applications: current status, classification and open issues. Telemat Inform 36:55–81
22. Chen HS, Jarrell JT, Carpenter KA, Cohen DS, Huang X (2019) Blockchain in healthcare: a patient-centered model. Biomed J Sci Tech Res 20(3):15017
23. Chen L, Lee W-K, Chang C-C, Choo K-KR, Zhang N (2019) Blockchain based searchable encryption for electronic health record sharing. Futur Gener Comput Syst 95:420–429
24. Christidis K, Devetsikiotis M (2016) Blockchains and smart contracts for the internet of things. IEEE Access 4:2292–2303
25. Chukwu E, Garg L (2020) A systematic review of blockchain in healthcare: frameworks, prototypes, and implementations. IEEE Access 8:21196–21214
26. Dagher GG, Mohler J, Milojkovic M, Marella PB (2018) Marella, Ancile: privacy-preserving framework for access control and interoperability of electronic health records using blockchain technology. Sustain Urban Areas 39:283–297
27. Dias J, Reis L, Ferreira H, Martins Â (2018) Blockchain for access control in e-health scenarios, arXiv: 180512267
28. Esposito C, De Santis A, Tortora G, Chang H, Choo K-KR (2018) Blockchain: a panacea for healthcare cloud-based data security and privacy?. IEEE Cloud Comput 5(1):31–37
29. Fan K, Wang S, Ren Y, Li H, Yang Y (2018) MedBlock: efficient and secure medical data sharing via blockchain. J Med Syst 42(8)
30. Farouk A, Alahmadi A, Ghose S, Mashatan A (2020) Blockchain platform for industrial healthcare: vision and future opportunities. Comput Commun 154:223–235
31. Fernández-Caramés TM, Fraga-Lamas P (2019) A review on the application of blockchain for the next generation of cybersecurity industry 4.0 smart factories. IEEE Access 7:45201–45218
32. Gao Z, Xu L, Chen L, Zhao X, Lu Y, Shi W (2018) CoC: a unified distributed ledger based supply chain management system. J Comput Sci Technol 33(2):237–248
33. Glaser F, Bezzenberger L (2015) Beyond cryptocurrencies a taxonomy of decentralized consensus systems. Book beyond Cryptocurrencies-A taxonomy of decentralized consensus systems, pp 1–18
34. Gordon WJ, Catalini C (2018) Blockchain technology for healthcare: facilitating the transition to patient-driven interoperability. Comput Struct Biotechnol J 16:224–230
35. Griggs KN, Ossipova O, Kohlios CP, Baccarini AN, Howson EA, Hayajneh T (2018) Healthcare blockchain system using smart contracts for secure automated remote patient monitoring. J Med Syst 42(7):1–7
36. Guardtime (2020) Solving real healthcare challenges with blockchain. https://guardtime.com/health
37. Guo R, Shi H, Zhao Q, Zheng D (2018) Secure attribute-based signature scheme with multiple authorities for blockchain in electronic health records systems. IEEE Access 6:11676–11686
38. Gupta S, Malhotra V, Singh SN (2020) Securing IoT-driven remote healthcare data through blockchain. In: Advances in data and information sciences. Springer, pp 47–56
39. Hasselgren A, Kralevska K, Gligoroski D, Pedersen SA, Faxvaag A (2020) Blockchain in healthcare and health sciences: a scoping review. Int J Med Inf 134. https://doi.org/10.1016/j.ijmedinf.2019.104040
40. Hathaliya J, Sharma P, Tanwar S, Gupta R (2019) Blockchain-based remote patient monitoring in healthcare 4.0. In 2019 IEEE 9th international conference on advanced computing (IACC), pp 87–91
41. Hölbl M, Kompara M, Kamišalić A, Nemec Zlatolas L (2018) A systematic review of the use of blockchain in healthcare. Symmetry 10(10):470
42. Hussein AF, ArunKumar N, Ramirez-Gonzalez G, Abdulhay E, JMRS T, de Albuquerque VHC (2018) A medical records managing and securing blockchain based system supported by a Genetic algorithm and discrete wavelet transform.Cogn Syst Res 52:1–11
43. Hussien HM, Yasin SM, Udzir SNI, Zaidan AA, Zaidan BB (2019) A systematic review for enabling of develop a blockchain technology in healthcare application: taxonomy, substantially analysis, motivations, challenges, recommendations and future direction. J Med Syst 43(10):1–35

44. Ichikawa D, Kashiyama M, Ueno T (2017) Tamper-resistant mobile health using blockchain technology. JMIR Mhealth Uhealth 5(7):e111. https://doi.org/10.2196/mhealth.7938
45. Identity Management Institute (2020) Blockchain for healthcare data security. https://identitymanagementinstitute.org/blockchain-for-healthcare-data-security/
46. Ismail L, Materwala H, Zeadally S (2019) Lightweight blockchain for healthcare. IEEE Access 7:149935–149951
47. Islam N, Faheem Y, Din IU, Talha M, Guizani M, Khalil M (2019) A blockchain-based fog computing framework for activity recognition as an application to e-healthcare services. Futur Gener Comput Syst 100:569–578
48. Islam SR, Kwak D, Kabir MH, Hossain M, Kwak KS (2015) The internet of things for health care: a comprehensive survey. IEEE Access 3:678–708
49. Jaoude JA, Saade RG (2019) Blockchain applications—usage in different domains. J IEEE Access 7:45360–45381
50. Jiang L, Chen L, Giannetsos T, Luo B, Liang K, Han J (2019) Toward practical privacy-preserving processing over encrypted data in iot: an assistive healthcare use case. IEEE Internet Things J 6(6):10177–10190
51. Kassab MH, DeFranco J, Malas T, Laplante P, Neto VVG (2019) Exploring research in blockchain for healthcare and a roadmap for the future. IEEE Trans Emerg Top Comput
52. Katuwal GJ, Pandey S, Hennessey M, Lamichhane B (2018) Applications of blockchain in healthcare: current landscape & challenges. arXiv:abs/1812.02776
53. Kleinaki AS, Mytis-Gkometh P, Drosatos G, Efraimidis PS, Kaldoudi E (2018) A blockchain based notarization service for biomedical knowledge retrieval. Comput Struct Biotechnol J 16:288–297
54. Koshechkin KA, Klimenko GS, Ryabkov IV, Kozhin PB (2018) Scope for the application of blockchain in the public healthcare of the Russian Federation. Procedia Comput Sci 126:1323–1328
55. Kotz D, Gunter CA, Kumar S, Weiner JP (2016) Privacy and security in mobile health: a research agenda. Computer 49(6):22–30
56. Liang X, Shetty, S, Zhao J, Bowden D, Li D, Liu J, Qing S, Liu D, Mitchell C, Chen L (2018) Towards decentralized accountability and self-sovereignty in healthcare systems, Springer, pp 387–398
57. Liang X, Zhao J, Shetty S, Liu J, Li D (2018) Integrating blockchain for data sharing and collaboration in mobile healthcare applications. Institute of Electrical and Electronics Engineers Inc.
58. Li H, Zhu L, Shen M, Gao F, Tao X, Liu S (2018) Blockchain-based data preservation system for medical data. J Med Syst 42(8):1–13
59. Mamoshina P, Ojomoko L, Yanovich Y, Ostrovski A, Botezatu A, Prikhodko P, Izumchenko E, Aliper A, Romantsov K, Zhebrak A, Obioma Ogu I, Zhavoronkov A (2018) Converging blockchain and next-generation artificial intelligence technologies to decentralize and accelerate biomedical research and healthcare. Oncotarget 9(5):5665–5690
60. McGhin T, Choo K-KR, Liu CZ, He D (2019) Blockchain in healthcare applications: research challenges and opportunities. J Netw Comput Appl 135:62–75
61. Mikula T, Jacobsen RH (2018) Identity and access management with blockchain in electronic healthcare records. In: The 21st Euromicro conference on digital system design (DSD), pp 29–31
62. Ministry of Social Affairs (2020) Factsheet: E-health in Estonia. https://na.eventscloud.com/file_uploads/c5da2a5e465f932e6debe55020e70899_E-health-factsheet.pdf
63. Monrat AA, Schelén O, Andersson K (2019) A survey of blockchain from the perspectives of applications, challenges, and opportunities. IEEE Access 7:117134–117151
64. Novikova K (2019) Top 5 blockchain projects in healthcare. https://digiforest.io/en/blog/blockchain-examples-in-healthcare
65. Nugent T, Upton D, Cimpoesu M (2016) Improving data transparency in clinical trials using blockchain smart contracts, F1000 Res

66. Omar I, Jayaraman R, Salah K, Simsekler M (2019) Exploiting ethereum smart contracts for clinical trial management. In: 2019 ACS 16th international conference on computer systems and applications (AICCSA). IEEE, Abu Dhabi, United Arab Emirates, pp 1–6
67. Panarello A, Tapas N, Merlino G, Longo F, Puliafito A (2018) Blockchain and IoT integration: a systematic survey 18(8)
68. Pandey P, Litoriya R (2020) Implementing healthcare services on a large scale: challenges and remedies based on blockchain technology. Health Policy Technol 9:69–78
69. Pareek S (2021) How blockchain will revolutionize logistics. Cloud Credential Council. https://www.cloudcredential.org/blog/how-blockchain-will-revolutionize-logistics/. Accessed 20 June 2021
70. Peterson K, Deeduvanu R, Kanjamala P, Boles K (2017) A blockchain-based approach to health information exchange networks
71. Pham HL, Tran TH, Nakashima Y (2018) A secure remote healthcare system for hospital using blockchain smart contract. In: 2018 IEEE globecom workshops (GC Wkshps), vol 1, pp 1–6
72. Radanović I, Likić R (2018) Opportunities for use of blockchain technology in medicine. Appl Health Econ Health Policy
73. Rahmadika S, Rhee K (2018) Blockchain technology for providing an architecture model of decentralized personal health information. International J Eng Bus Manag
74. Rahman MA, Hossain MS, Hassanain E, Rashid M, Barnes S (2018) Spatial blockchain-based secure mass screening framework for children with dyslexia. IEEE Access
75. Ray PP, Dash D, Salah K, Kumar N (2020) Blockchain for IoT-based healthcare: background, consensus, platforms, and use cases. IEEE Syst J 1–10
76. Reyna A, Martín C, Chen J, Soler E, Díaz M (2018) On blockchain and its integration with IoT. challenges and opportunities. Futur Gener Comput Syst 88:173–190
77. Roehrs A, da Costa CA, da Rosa Righi R (2017) OmniPHR: a distributed architecture model to integrate personal health records. J Biomed Inform 71:70–81
78. Sahi MA, Abbas H, Saleem K, Yang X, Derhab A, Orgun M, Iqbal W, Rashid I, Yaseen A (2017) Privacy preservation in e-healthcare environments: state of the art and future directions. IEEE Access 6:464–478
79. Salah K, Rehman MHU, Nizamuddin N, Al-Fuqaha A (2019) Blockchain for AI: review and open research challenges. IEEE Access 7:10127–10149
80. Sengupta J, Ruj S, Bit SD (2020) A comprehensive survey on attacks, security issues and blockchain solutions for IOT and IIOT. J Netw Comput Appl 149:102481
81. Shahnaz A, Qamar U, Khalid A (2019) Using blockchain for electronic health records. IEEE Access 7:147782–147795
82. Shan J, Jiannong C, Hanqing W, Yanni Y, Mingyu M, Jianfei H (2018) BlocHIE: a blockchain-based platform for healthcare information exchange. In: IEEE international conference on smart computing (SMARTCOMP). IEEE Computer Society, Los Alamitos, CA, USA, pp 18–20
83. SHIVOM (2018) Blockchain can be the catalyst for a revolution in precision medicine. https://medium.com/projectshivom/blockchain-can-be-the-catalyst-for-a-revolution-in-precision-medicine-d55e1e810262
84. Shrestha AK, Vassileva J (2016) Towards decentralized data storage in general cloud platform for meta-products. In: Proceedings of the international conference on big data and advanced wireless technologies, pp 1–7
85. Smith SM, Khovratovich D (2016) Identity system essentials. Evemyrn
86. Sun J, Yan J, Zhang KZK (2016) Blockchain-based sharing services: what blockchain technology can contribute to smart cities? Financ Innov 2(26):1–9
87. Syed TA, Alzahrani A, Jan S, Siddiqui MS, Nadeem A, Alghamdi T (2019) A comparative analysis of blockchain architecture and its applications: problems and recommendations. IEEE Access 7:176838–176869
88. Talal M, Zaidan A, Zaidan B, Albahri A, Alamoodi A, Albahri O, Alsalem M, Lim C, Tan KL, Shir W (2019) Smart home-based IOT for real-time and secure remote health monitoring

of triage and priority system using body sensors: multi-driven systematic review. J Med Syst 43(3):615–620
89. Tandon A, Dhir A, Islam N, Mäntymäki M (2020) Blockchain in healthcare: a systematic literature review, synthesizing framework and future research agenda. Comput Ind 122:103290
90. Tanwar S, Parekh K, Evans R (2020) Blockchain-based electronic healthcare record system for healthcare 4.0 applications. J Inf Secur Appl 50:102407
91. Tao H, Bhuiyan MZA, Abdalla AN, Hassan MM, Zain JM, Hayajneh T (2018) Secured data collection with hardware-based ciphers for Iot-based healthcare. IEEE Internet Things J 6(1):410–420
92. Uddin MA, Stranieri A, Gondal I, Balasubramanian V (2018) Continuous patient monitoring with a patient centric agent: a block architecture. IEEE Access 6:32700–32726
93. Williams-Grut O (2016) Estonia is using the technology behind bitcoin to secure 1 million health records. https://www.businessinsider.com.au/guardtime-estonian-health-records-industrial-blockchain-bitcoin-2016-3?r=US&IR=T
94. Xia Q, Sifah EB, Asamoah KO, Gao J, Du X, Guizani M (2017) MeDShare: trust-less medical data sharing among cloud service providers via blockchain. IEEE Access 5:14757–14767
95. Xia Q, Sifah EB, Smahi A, Amofa S, Zhang X (2017) BBDS: blockchain-based data sharing for electronic medical records in cloud environments. Inf (Switzerland) 8(2)
96. Xie J, Tang H, Huang T, Yu FR, Xie R, Liu J, Liu Y (2019) A survey of blockchain technology applied to smart cities: research issues and challenges. IEEE Commun Surv Tutor 21(3):2794–2830
97. Yaqoob I, Salah K, Jayaraman R, Al-Hammadi Y (2021) Blockchain for healthcare data management: opportunities, challenges, and future recommendations. Neural Comput Appl J 1–19
98. Yu H, Yang Z, Sinnott RO (2019) Decentralized big data auditing for smart city environments leveraging blockchain technology. IEEE Access 7:6288–6296
99. Yue X, Wang H, Jin D, Li M, Jiang W (2016) Healthcare data gateways: found healthcare intelligence on blockchain with novel privacy risk control. J Med Syst 40(10):218
100. Zhang A, Lin X (2018) Towards secure and privacy-preserving data sharing in e-health systems via consortium blockchain. J Med Syst 42(8)
101. Zhang M, Ji Y (2018) Blockchain for healthcare records: a data perspective. Peer J Preprints 6:e26942v1
102. Zhang P, White J, Schmidt D, Lenz G (2017) Applying software patterns to address—Interoperability in blockchain-based healthcare apps, arXiv:170603700
103. Zhang P, White J, Schmidt DC, Lenz G, Rosenbloom ST (2018) FHIR chain: applying blockchain to securely and scalable share clinical data. Comput Struct Biotechnol J 16:267–278
104. Zhang X, Poslad S (2018) Blockchain support for flexible queries with granular access control to electronic medical records (EMR). Institute of Electrical and Electronics Engineers Inc.
105. Zheng Z, Xie S, Dai H-N, Chen W, Chen X, Weng J, Imran M (2020) An overview on smart contracts: challenges, advances and platforms. Futur Gener Comput Syst 105:475–491
106. Zheng Z, Xie S, Dai H-N, Chen X, Wang H (2016) Blockchain challenges and opportunities: a survey. Int J Web Grid Serv 1–24
107. Zhou L, Wang L, Sun Y (2018) MIStore: a blockchain-based medical insurance storage system. J Med Syst 42(8):1–17
108. Zhu H, Wu CK, Koo CH, Tsang YT, Liu Y, Chi HR, Tsang K-F (2019) Smart healthcare in the era of internet-of-things. IEEE Consum Electron Mag 8(5):26–30

Blockchain as a Service: A Holistic Approach to Traceability in the Circular Economy

Benítez-Martínez Francisco Luis, Nuñez-Cacho-Utrilla Pedro Víctor, Molina-Moreno Valentín, and Romero-Frías Esteban

Abstract Today blockchain technology provides us with a formidable tool in the struggle to trace economic resources, especially in the context of the circular economy. The circular economy has been proposed as a key element in the transformation of production models in the context of European post–covid-19 recovery plans with particular reference to the Next Generation EU instrument. It is also a fundamental part of the European Green Deal. All of which comes under the umbrella of the United Nations' Sustainable Development Goals and the 2030 Agenda. The circular economy lays the foundations for the promotion of a new production and consumption model in which the value of products, materials, and resources remains within the economy for as long as possible, minimizing the generation of waste. This gives rise to a series of processes in which resource traceability is a key factor in preserving process integrity and guaranteeing process authenticity to the State, citizens, and companies. In this context, blockchain technology can provide solutions that are aligned with the 2030 Agenda. This technology facilitates the procedures and

B.-M. Francisco Luis
FIDESOL, BIC Building, Avda. Innovación, s/n. Health Technology Park (PTS), 18100 Armilla, Granada, Spain
e-mail: flbenitez@fidesol.org

B.-M. Francisco Luis · R.-F. Esteban (✉)
MediaLab, University of Granada, Avda. Madrid, s/n, Espacio V Centenario, 18071 Granada, Spain
e-mail: erf@ugr.es

N.-C.-U. Pedro Víctor
Polytechnic School of Linares, University of Jaen, Ronda Sur s/n, 23700 Linares (Jaen), Spain
e-mail: pnunez@ujaen.es

M.-M. Valentín
Department of Management-1, Faculty of Economics and Business, University of Granada, Campus Cartuja, s/n, 18071 Granada, Spain
e-mail: vmolina2@go.ugr.es

R.-F. Esteban
Department of Accounting and Finance, Faculty of Economics and Business, University of Granada, Campus Cartuja, s/n, 18071 Granada, Spain

S. S. Muthu (ed.), *Blockchain Technologies for Sustainability*,
Environmental Footprints and Eco-design of Products and Processes,
https://doi.org/10.1007/978-981-16-6301-7_6

processes of logistics and of IoT sensor records via smart contracts through intrinsic properties that include timeproof sealing and data record immutability. In the present chapter, we describe the technological advantages that blockchain technology offers the circular economy. Sustainability is the cornerstone of blockchain models within the framework of the 2030 Agenda, so energy pollution in transactions or mining should be avoided. If we are able to overcome current environmental deterrents, distributed ledger technologies should represent a powerful tool in circular economy projects. In this chapter, we would hope to contribute to the debate on future paths towards sustainability. Specifically, we will describe how Blockchain as a Service-based traceability platform could be introduced into the circular economy while guaranteeing their straightforward but highly effective deployment in, for instance, the agrifood sector, at only minimal cost to SMEs. The underlying idea is based on finding blockchain solutions aligned with Sustainable Development Goals in order to ensure that the principal objectives and philosophy of the circular economy are upheld.

Keywords Blockchain · Distributed ledger technologies · Circular economy · Traceability · Sustainable development goals · Agenda 2030 · Blockchain as a service · Sustainability

1 Introduction

The emergence of distributed ledger technologies (DLTs), with the blockchain at the forefront, provides an opportunity to establish a new framework for processes in circular economy (CE) projects. This chapter offers both an overview of the CE as a concept and of the challenges it faces, and a novel solution to the issue of traceability within the CE by using a Blockchain as a Service (BaaS) model that relies on a cutting-edge DLT which, unlike classic blockchain and Ethereum approaches, guarantees sustainability. A key concept of our model is that of token and "tokenization" which, beyond the use of technology as a support for cryptocurrencies, facilitates digitizing and operating with any other type of asset.

In Sect. 2, we introduce the CE, address its principal concepts, and look at its evolution through a variety of configurations proposed by different authors. Section 3 includes our BaaS proposal, which uses a Neural Distributed Network (NDL)-based solution that overcomes the disadvantages of blockchain and Ethereum—in terms of sustainability due to their high energy consumption—and offers additional advantages. Finally, in Sect. 4, we conclude by reflecting on the implications and future challenges represented by the employment of DLTs specifically, and of emerging digital technologies in general, in order to improve CE models and fulfill the 2030 Agenda.

2 New Paradigms: The Circular Economy and Blockchain Technology

2.1 Introduction to the Circular Economy

Today, terms such as climate change, greenhouse gas emissions, pollution, environmental toxicity, or loss of biodiversity are widely used to describe part of the reality with which *Homo sapiens* engages. This reality is based on what is called a linear economy or a traditional economy model and is characterized by extracting, transforming, and generating waste—as such it has made itself unsustainable [13, 33]. The inefficiency of the linear model is mainly due to the negative externalities it generates—destabilizing and endangering the economic, environmental, and social sustainability of our ecosystems—and how they affect natural ecosystems [10–12, 14, 38, 45, 50]. All this is a challenge to the survival of organizations, the global economy, and, indeed, the planet itself—as is reflected in the UN's Sustainable Development Goals (SDGs). Consequently, we need to respond to this inefficiency and move towards a new paradigm that is based on sustainable sociotechnical systems [24, 42].

2.1.1 The Circular Economy as an Emerging Discipline

The circular economy has materialized against this backcloth. The origins of the paradigm are deep-rooted and did not come into being on a specific date or through a specific author. As early as 1965, Kenneth E. Boulding pointed out that humanity had for centuries thought that resources were unlimited. In the context of a sparsely populated planet, this was quite conceivably the case. However, the evolution of the Earth's population has converted ours into a finite, limited, crowded planet. That is why Boulding proposed we think of the earth as if it were a spaceship, from which we cannot throw out waste. He introduced the idea of "stable circular flow", necessary for the medium-term survival of the planet.

Subsequently, the so-called Performance Economy appeared. In a report to the European Commission, Stahel and Reday-Mulvey [44] introduced the idea of substituting the use of energy for that of labor. We should recall that in the 1970s energy prices soared and unemployment levels grew significantly. In their report, the authors argued that more manpower and fewer resources were needed to renovate existing buildings rather than erect new ones. They included the vision of an economy in cycles or loops (a circular economy) that, in addition to having a positive impact on job creation and economic competitiveness, would save resources and reduce waste generation. These principles were valid for any stock or capital, from mobile phones to arable land or our cultural heritage. Stahel is credited with having invented the expression "[from] Cradle to Cradle" (C2C). This links with the idea of regenerative design, which developed from the principles of agriculture and was initially applied to architecture. This idea of regeneration is one of the foundations of the

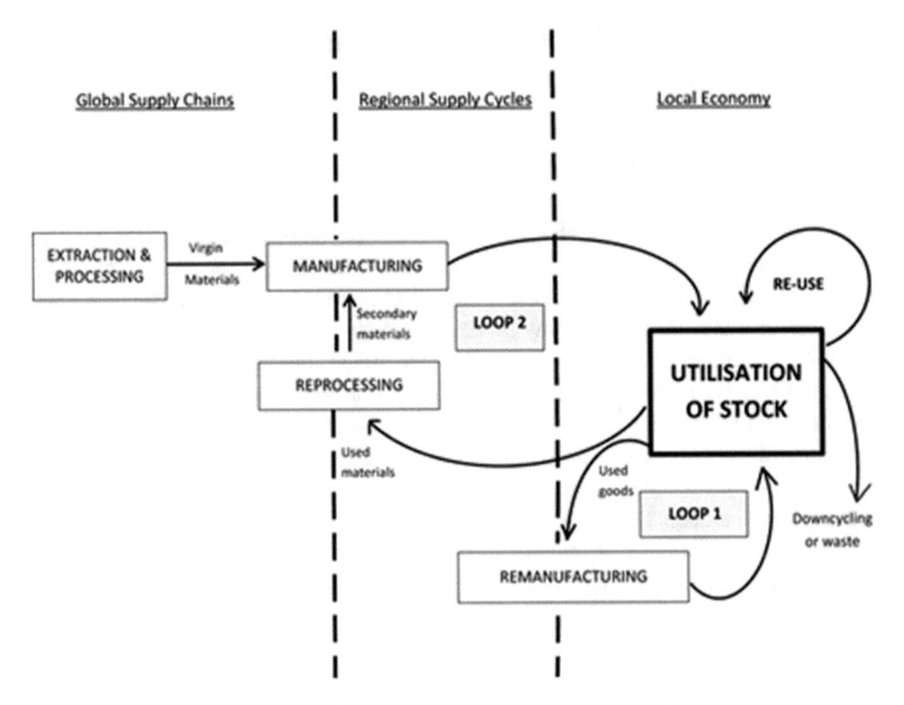

Fig. 1 The basic loops of a circular economy. *Source* Stahel and Clift [43]

CE framework, developed thanks to Lyle, McDonough, Braungart, and Stahel. As early as 1980, CE as a term was used to describe a system of interaction between the economy and the environment [39] (Fig. 1).

The C2C philosophy was initially mentioned in *Cradle-to-Cradle: Remaking the Way We Make Things* [26]. The authors proposed that products should be made following C2C principles, designed and manufactured to avoid environmental contamination in their manufacture and throughout their useful life.

Production based on C2C involves a circular industrial system in which all materials are used indefinitely, and which is fully supported in the reverse supply chain. At the end of their useful life, materials become a primary resource—theoretically, with no loss of quality—to manufacture the same product or a different one; the general process can be considered an "ascending cycle" [9].

The C2C framework highlights the importance of designing products that have a positive impact on their environment and negative impacts that are reduced through efficiency. The C2C design views the safe and productive processes of nature's 'biological metabolism' as a model for the development of a 'technical metabolic flow' of industrial materials. Product components can be designed for continuous recovery and reuse as biological and technical nutrients in these metabolisms. The C2C framework eliminates the traditional concept of waste since waste equates with nutrients. In addition, systems that collect and recover the value of these materials after their use need to be created and used. The need to control energy and water input is also addressed. One of the key principles of the CE foregrounds energy: only renewable energy sources should be used and networks should be decentralized. Water use must promote healthy ecosystems and respect its effects at the local level.

Another school of thought on which the CE is based in the Theory of Industrial Ecology. This presents industry as an ecosystem [19] and applies the principles of the Theory of Human Ecology to study the interaction between industries and the environment and their interdependence, creating a basic ecological framework for decision-making [36].

The Theory of Industrial Ecology presents a synthesis of ecology-based assumptions, concepts, and propositions drawn from several disciplines and from general systems theory. It enables us to describe and explain interactions that occur within industries and their transactions with the environment. The perspective is scientifically based but its principles, methods, and results are applied to everyday activities [7]. The theory refers to the "creation, use and management of resources for adaptation, human development and the sustainability of environments", focusing on interactions between industry and the environment.

2.1.2 The Three Starting Points of the Circular Economy

Three principles underlie the CE's starting point. Firstly, any given industry interacts with its environment creating an ecosystem in which, according to systems theory, the parts and whole are interdependent. Interactions occurring within this ecosystem are guided by physical, biological, and social laws and continually draw matter and energy from the ecosystem. Secondly, nutritional, physical, biological, economic, and psychosocial maintenance functions are performed within this ecosystem. Hence, the theory holds immense value when examining these functions at different levels and their interrelationships over time. Thirdly, industry and resources are interdependent. Therefore, ecological wealth depends on the decisions and actions of countries and industries. The welfare of a given industry cannot be separated from the welfare of the entire ecosystem [6]. So, we must establish a balance between the demands of the ecosystem and those of individuals.

Biomimicry

Janine Benyus defines this approach as a new discipline that studies the best ideas of nature and then imitates these designs and processes to solve human problems.[1] She thinks of biomimicry as Nature-inspired innovation. Biomimicry refers to the philosophy and interdisciplinary design approaches that take Nature as a model when facing the challenges of sustainable development (social, environmental, and economic). Within this framework, biomimetics (including 'biomimetic design') entails the interdisciplinary cooperation of biology and technology or other innovative fields, in order to solve practical problems by analyzing functions of biological

---

[1] http://www.biomimicryinstitute.org.

systems, their abstraction in models, and the transfer and application of these models to the solution.

Blue economy

Driven by Gunter [37], a Belgian businessman and former CEO of Ecover, the Blue economy is an open source movement that brings together a series of case studies compiled in a report of the same name delivered to the Club of Rome. As the official manifesto says, "using the resources available in cascade systems, (…) the waste of a product becomes the input to create a new cash flow." The main idea is to copy the behavior of natural ecosystems in the way that they consume local products and do not generate waste. For Pauli, all waste generated should be reintroduced into the production process as new raw materials, exactly as happens in Nature. The use of waste for other processes would also benefit the economy by generating new jobs.

The final pillar of the CE is so-called "natural capital", which refers to the world's reserves of natural assets including soil, air, water, and all living beings. In *Natural Capitalism: Creating the Next Industrial Revolution* [18], Paul Hawken, Amory Lovins, and L. Hunter Lovins describe an economy in which business and environmental interests overlap, recognizing the interdependencies that exist between the production and use of man-made capital and natural capital flows.

2.1.3 The Concept of the Circular Economy

The CE is a restorative and regenerative industrial economy by intention and by design [23, 25]. It is designed to depend on renewable energy, minimizing the use of fossil fuels. In general, it eliminates or mitigates the use of toxic chemicals and waste through a design that stipulates minimal resource use in manufacturing and a reduction in the production-energy balance, water footprint, and carbon footprint. A company can be considered to produce according to the CE paradigm when its process is regenerative with respect to input and it has a low environmental impact in terms of greenhouse gas emissions and water footprint. Production under the CE paradigm seeks to optimize the process and minimize the negative externalities generated [27, 34]. These concepts are reinforced by the following principles.

Product as a service. The use of the products is sold but the material is not; the consumer simply uses the product and the supplier is responsible for recycling the material. Customers can purchase the use as a service and when the product becomes obsolete, it is withdrawn from the market and renewed.

Reuse is a symbol of good management. The principle of the 3Rs (reduce, reuse and recycle) contributes to reducing the pressure on the stock of global resources [40]. Companies must develop new production models by broadening this philosophy to include the "10Rs" (reduce, reuse, reject, rethink, redistribute, repair, restore, reuse, recycle and recover).

The CE is based on life cycle biomimetics, with technology mimicking the behavior of biological cycles. It is all about imitating the natural biological recycling process in the use of industrial materials [4]. This is how the concept of biological and technological nutrients arises. Biological nutrients are materials that can be renewed without a human process. Biological nutrients are recycled through biochemical processes such as conversion into biomass, anaerobic digestion, in which organic matter kills microorganisms and produces biogas; and composting, applied to soil restoration. Technological nutrients require that human action be reincorporated into the system as 'food' for another process [35].

The CE model minimizes the negative externalities generated. Krugman [21] pointed to the link between market exchanges and negative externalities (costs incurred by third parties when the first two—buyers and sellers—conduct economic transactions). Negative externalities can manifest themselves as costs related to economic, environmental, and/or social sustainability for third parties and for society in general. Laczniak [22] suggests that negative externalities represent the hidden costs of market exchanges and that we remain ignorant of their true costs unless these are identified and quantified. The most important cost of certain market exchanges may not be the price paid for the products but the damage to the natural environment caused by toxic effluents derived from the extraction, production, distribution, consumption, and post-consumption stages.

Circular economy fundamentals have been strengthened by reflections and contributions drawn from science and aimed at improving the efficiency of resources in order to balance the relationship between the economy, the environment, and society [15, 16, 29]. Analysis of the CE is increasingly detailed as its effects on different economic sectors are studied. This is demonstrated by the generation of new publications in scientific journals and in reports and case studies [1, 2, 28, 32, 49] (EMF [10]; Yap [48]).

2.1.4 The Potential Use of Blockchain in the Circular Economy

The application of blockchain in the CE is an emerging field of study given that this technology has an enormous potential to overcome the risks inherent to CE projects that contribute to sustainable development and to fulfilling the 2030 Agenda's SDGs. Böckel et al. [5] stated that certain key factors must be taken into account when choosing a DLT. They focused on DLT's technical properties and determined that further research should address the definition of clear terminology and the benefits of DLT use. They also specified that trust and data verification are major potential benefits but stressed that the possible benefits and challenges of blockchain technologies needed closer examination.

We cannot forget that blockchain technology is associated with decentralization, anonymity, proof tampering, and auditability. All of these can be used in many use cases: in supply chain management, emissions trading, agricultural processes, and food and energy traceability, for instance. These applications are closely linked to CE principles, enabling any kind of CE project to avoid fraud and secure data and management information and create a new model of processes that are transparent to end-users.

Classic blockchain and Ethereum present serious problems in terms of sustainability due to their high energy consumption [46]. Moreover [41], showed how the employment of blockchains that use the PoW (Proof of Work) consensus mechanism is unsuitable for CE projects. In the next section, we propose a novel solution to CE traceability by using a BaaS model that relies on cutting-edge DLT that guarantees sustainability.

3 A Blockchain Integration Model for the Circular Economy. A BaaS Approach

3.1 Blockchain and the Circular Economy

Blockchain and the CE are two emerging concepts with potentially beneficial interrelations. The CE addresses the minimization of negative externalities generated by organizations. These negative externalities are derived from the processes of creating and using purchased products, and from their end of life. In this context, since blockchain technology is a protocol that shares and updates information linking accounting books or databases in a decentralized, peer-to-peer, open-access network, it can contribute to the formalization of the CE in various ways and, thus, facilitate the implementation of its regenerative, restorative vision [20, 29, 47].

In situations in which blockchain has been applied to sustainability, the issue most commonly studied has been supply chain traceability. This is undoubtedly a key issue since blockchain technology would reliably and safely guarantee the circularity of materials that the CE requires. As company sustainability strategies evolve and sustainability reports demand a high volume of indicators, the management of reliable, safe data on issues such as water and energy consumption, chemical substance use, materials, or the plastic footprint constitutes an imperative [8]. Records of these and other sustainability data can be guaranteed by blockchain.

This technology provides a sound, distributed, interconnected system that secures and tests any organizations' transition to the CE [47]. It can ensure that data is stored and updated in a secure, tamper-proof, irreversible manner.

Therefore, we believe that the use of blockchain technology can drive and accelerate organizational transition to the CE, with reduced transaction costs and optimal levels of performance and communication—all of which are much needed throughout the supply chain [5]. The technology will provide any link in the supply chain with detailed information about materials, emission indicators, and so on while improving data confidentiality [20] and reducing the carbon footprint [47].

While recording all the information is important for vendors in the supply chain, blockchain technology's importance to consumers cannot be ignored. When consumers focus on sustainable purchasing decisions, they demand validated sustainable practices to give them confidence in their choice of product. Hence, transparent and clearly defined sustainability policies will act as an incentive for purchase decisions [35]. In this context, blockchain technology will communicate the exact origin of materials and inform consumers as to whether or not a product has really been manufactured according to EC-defined parameters.

Furthermore, the technology will also be of use in the treatment of waste. Thanks to records provided by the blockchain, individuals and recycling companies will know the precise nature of the products they have to recover, how they should do so, and how those products will re-enter the supply chain—whether through reuse, remanufacturing, recycling, or whatever.

One of the most important issues to take into account when integrating blockchain technology into the CE is the sustainability of the option chosen. Analysis of early blockchain technology has shown that bitcoin totally contradicts the principles of the CE. The high energy needs entailed in mining the cryptocurrency and in a governance system based on a highly speculative design make it inadvisable. Moreover, it totally contradicts the 2030 Agenda's SDGs. The case of Ethereum is similar. Although this blockchain network introduces smart contracts as a noteworthy, intelligent solution to simplify processes and improve efficiency, it still uses consensus transaction systems—like PoW or PoS (Proof of Stake)—which constitute separate networks that move away from the paths of sustainability due to the energy they invest in mining tokens. Nonetheless, it is an interesting starting point from which to establish designs and applications that help develop CE projects that ensure their traceability and transparency. This public network offers an interoperability framework that can connect current blockchain solutions with those of the future. Blockchain is especially relevant in the context of the 2030 Agenda since it surpasses the technological framework of bitcoin and Ethereum [17].

Therefore, finding sustainable solutions with the use of blockchain applications is essential to CE projects. It will offer easy access for all and minimal energy consumption in frameworks that deploy blockchain dApps.

Nowadays, most organizations are keen to adopt the concept of CE. Natural resources are generally accepted to be finite and irreplaceable, which is inconsistent with economic growth; moreover, public opinion is highly concerned as citizens and consumers [29] need new tools to track the implications of their actions in a perilous scenario in which the post-pandemic crisis and climate change are the major challenges facing society. So, a blockchain framework can have a positive impact on the main goals of a CE and on implications for the ethics agenda [47]. A new socially responsible approach could emerge as a consequence of this technology-based intervention.

The covid-19 pandemic has shown us that business and society are fragile and lack resilience even in the face of the typical *pre*-pandemic patterns of production and consumption—particularly with respect to supply chains and logistics. More than ever we need to build economies that are more sustainable, transparent, and trustworthy. Digitization is an opportunity to establish a new framework for CE projects that include Industry 4.0 technologies with a special focus on blockchain [30].

One of the most interesting aspects is the use of the CE objectives as intrinsic assets in the blockchain solution. This makes tokenization of the projects an opportunity to enhance close cooperation between market and consumers in order to fulfill CE principles. In this context ([31]: 6), pointed to the incentives of tokenization in order to offer an ecosystem to cooperate and compete to construct strong CE environments, creating a framework for competition and using tokens to build them.

3.2 The BaaS Approach

Since it meets the CE's requirements for sustainability, a BaaS approach is one of the best options in order to build scalable, interoperable solutions. From regulators to end-users, a substantial number of stakeholders actively participate in BaaS, playing a great number of roles and exhibiting a variety of interests. These actors apply different capabilities in order to deploy blockchain solutions, regardless of the DLT being used.

In order to deploy easy, scalable solutions, we can use dApps based in a BaaS context. We define a BaaS perspective as being one in which we offer a cloud-based solution for users to build their own dApps and their own blockchains. These would include applications and smart contracts and defy the current mainstream framework under the umbrella of the Ethereum platform.

Open APIs of any kind of DLT together with a new and greener approach is essential. Neither Bitcoin nor Ethereum is eco-friendly solutions and both contradict the SDGs with regard to energy consumption and operative governance models. Sustainable Development Goals 16 and 17 demand a new governance model for future DLTs. Therefore, a new generation of DLTs will need to provide an accurate answer to the problems of the currently best-known blockchain platforms.

Federated DLT clouds together with cloud providers facilitate the easy, cheap deployment of dApps, cutting consumption of electricity and simplifying dApp adoption, both in the market and for end-users. Furthermore, as propose, this approach can facilitate the tokenization of IoT devices, as we mentioned earlier. Merging a BaaS perspective with IoT tokenization (a BIoT approach) is one of the most striking challenges in securing the traceability of CE project processes.

An optimal CE solution demands a blockchain platform with:

(a) a set of specific design data management patterns and a clear and affordable governance model for all stakeholders.
(b) a set of smart contracts built (or adapted from existing ones) for a CE environment.
(c) a ledger scheme open to new approaches and consensus models (not only protocols). If we introduce a federated consortium, the blockchain's construction would differ from that of current frameworks.
(d) The scalability and feasibility of the proposed solutions need to be adapted to the tokenization model and guarantee the quality of the information contained

in these tokens. Solutions prepared for an on-chain scheme are safer than those using off-chain records.

(e) The interoperability of the platform used is essential. Its capacity to exchange data and tokens with other blockchains will offer a clear opportunity to reach the market and be considered by end-users. A "killer-dApp" used in a CE project constitutes a genuine opportunity to offer blockchain technology to consumers.

3.3 A Specific, Neural Distributed Network-Based Solution

Given our earlier account of the NDL solution, we recommend the following, which is based on a tokenizable model for CE projects. Within this kind of DLT framework—called RETIS—a two layer ledger enhances the security of platform-issued tokens.

This ledger of ledgers is called ARCA. It acts on two levels; on the first, it guarantees the transmission of the data contained in the tokens, distributing this to all participating nodes. On the second level—which controls the governance of ledger management and of the transactions that occur in it, with specific hashes for their issuance—all nodes are synchronized.

This system verifies the transactions, unifying the entire final ledger with a cascade-of-nodes system. To do so, it uses two types of nodes: so-called "slaves", which are established by the light nodes (those deployed in a cloud); and "master" nodes, which are deployed within the platform system with dedicated hardware. The cascade-of-nodes system is designed to perform double verification of the data and/or smart contracts to be transacted. The slave nodes are in charge of verifying all orders from the master node before including them in the corresponding block. This system issues tokens with an average of 150000 transactions per second, use no type of payment and does not involve high energy costs because of its internal governance.

We could say that ARCA is a token store, based on a specifically designed database, which can protect assets from any physical medium in real environments. It acts as a virtual safe box with two different private keys to access the information stored in each token contained within it. As well as facilitating massive transactions, it can be deployed on all types of devices (smartphones, Raspberry, tablets), and can be used with multiple technologies (Doker, Kubernetes, etc.). By virtue of its dual-tier architecture, ARCA seamlessly integrates with Hyperledger for guaranteed interoperability. Hence, the tokens issued can be exported or imported from other ledgers.

A 3-D network, RETIS accumulates nodes in different blockchains within the framework. It facilitates block building in each of the blockchains on the platform. This ensures vertical and horizontal scalability (permitting thousands of transactions per second in RETIS) that forms a 3D mesh to deal with all data from all operators in a CE project. This process was defined by [3]. Moreover, the platform includes EQUO, a framework that acts as a tool to develop native dApps.

All of these features make this kind of DLT a transparent, green option for use in CE projects.

4 Implications and Challenges

As the CE is a multidisciplinary paradigm in progress, it needs contributions from the scientific fields that surround it. Contributions from technology are also necessary and welcome. The blockchain in particular can be a validation tool for CE-related issues of sustainability. However, we would stress that it must also be supported by sustainable mining, with renewable energy and the reuse of materials and equipment, to avoid increasing companies' carbon footprints. Once this obstacle has been overcome, we consider that blockchain technology will offer the CE a wide range of alternative applications, especially with regard to the measurement, certification, and recording of sustainability indicators, measurements, parameters, footprints, and so on.

We are currently at too early a stage to fully assess the potential that combing emergent digital technologies could represent for a more sustainable economy, a true circular economy. Both CE and blockchain technology have emerged in the last decade, and they are not alone. The current digital revolution affects emerging technologies like Artificial Intelligence, robotics, the IoT, and quantum computing. The creative, experimental combination of all of these will enable us to generate more efficient solutions and tackle new, more complex problems than we are currently capable of handling. Energy sustainability in the use of these technologies is essential for them to form part of a general CE framework. Similarly, the consumption of minerals and other resources in the manufacture of these systems poses significant challenges to the CE itself. While it is true that digitization enables us to reduce certain types of consumption—of paper, for example—it sharply increases others, generating environmental and social problems.

Digital technology in general—including blockchain—and the CE will go hand in hand for decades to come. We predict significant successes together with challenges that are even more complex than those we currently face, given that social expectations of these systems are gradually growing. In the future, one very important challenge will be to determine how to merge blockchain and Artificial Intelligence solutions to achieve more efficient, sustainable platforms and tools to manage CE projects. Another will be how to deploy simple, easily-used, state-of-the-art blockchains (DLTs) in specific projects, within a scalable, interoperable framework, which provides all stakeholders with easy access.

As we have discussed in this chapter, without doubt, DLTs, particularly those derived from the new technological approaches we propose here, currently provide sustainable answers to resolve problems of traceability in supply chains and recycling. Our goal is to continue to develop these models and to incorporate other technologies that, in combination with DLTs, provide even more comprehensive solutions.

References

1. Andersen MS (2007) An introductory note on the environmental economics of the circular economy. Sustain Sci 2(1):133–140. https://doi.org/10.1007/s11625-006-0013-6
2. Argudo-García JJ, Molina-Moreno V, Leyva-Díaz JC (2017) Valorization of sludge from drinking water treatment plants. A commitment to circular economy and sustainability. Dyna 92(1):71–75
3. Benítez-Martínez FL, Hurtado-Torres MV, Romero-Frías E (2021) A neural blockchain for a tokenizable e-Participation model. Neurocomputing 423:703–712. https://doi.org/10.1016/j.neucom.2020.03.116
4. Blériot J (2013) Cradle to cradle-products, but also systems. Ellen Macarthur Foundation. http://goo.gl/K87JHB
5. Böckel A, Nuzum AK, Weissbrod I (2020) Blockchain for the circular economy: analysis of the research-practice gap. Sustain Prod Consum 25:525–539. https://doi.org/10.1016/j.spc.2020.12.006
6. Brown L, Flavin C, Postel S (1989) A world at risk. In: Brown L, Flavin C, Postel S, Starke L (eds) State of the world: a Worldwatch Institute report on progress toward a sustainable society. New York, Norton, pp 3–20
7. Bubolz MM, Sontag MS (1993) Human ecology theory. In: Boss P, Doherty WJ, LaRossa R, Schumm WR, Steinmetz SK (eds) Sourcebook of family theories and methods: a contextual approach. Springer Science & Business Media, New York
8. Chidepatil A, Bindra P, Kulkarni D, Qazi M, Kshirsagar M, Sankaran K (2020) From trash to cash: how blockchain and multi-sensor-driven artificial intelligence can transform circular economy of plastic waste? Adm Sci 10(2):23. https://doi.org/10.3390/admsci10020023
9. Contreras-Lisperguer R, Muñoz-Cerón E, Aguilera J (2017) Cradle-to-cradle approach in the life cycle of silicon solar photovoltaic panels. J Clean Prod 168:51–59. https://doi.org/10.1016/j.jclepro.2017.08.206
10. Ellen MacArthur Foundation (EMF) (2013) Towards the circular economy 1: Economic and business rationale for an accelerated transition. https://www.ellenmacarthurfoundation.org/es/economia-circular/concepto (accessed 21 february 2021)
11. European Commission (2014a) MEMO, questions and answers on the commission communication “Towards a Circular Economy” and the waste targets review. https://ec.europa.eu/commission/presscorner/detail/en/MEMO_14_450
12. European Commission (2014b). Communication from the Commission to the European Parliament, the Council, the European Economic and Social Committee and the Committee of the Regions. Towards a Circular Economy: a Zero Waste Programme for Europe. COM, 398
13. Frosch RA, Gallopoulos NE (1989) Strategies for manufacturing. Sci Am 261(3):144–153. https://www.jstor.org/stable/24987406
14. Geng Y, Fu J, Sarkis J, Xue B (2012) Towards a national circular economy indicator system in China: an evaluation and critical analysis. J Clean Prod 23(1):216–224. https://doi.org/10.1016/j.jclepro.2011.07.005
15. Ghisellini P, Cialani C, Ulgiati S (2016) A review on circular economy: the expected transition to a balanced interplay of environmental and economic systems. J Clean Prod 114:11–32. https://doi.org/10.1016/j.jclepro.2015.09.007
16. Giampietro M (2019) On the circular bioeconomy and decoupling: implications for sustainable growth. Ecol Econ 162:143–156. https://doi.org/10.1016/j.ecolecon.2019.05.001
17. Gomez-Trujillo AM, Velez-Ocampo J, Gonzalez-Perez MA (2021) Trust, Transparency, and technology: blockchain and its relevance in the context of the 2030 Agenda. In: The Palgrave handbook of corporate sustainability in the digital Era. Palgrave Macmillan, Cham, pp 561–580. https://doi.org/10.1007/978-3-030-42412-1_28
18. Hawken P, Lovins AB, Lovins LH (2013) Natural capitalism: the next industrial revolution. Routledge
19. Hook N, Paolucci B (1970) The family as an ecosystem. J Home Econ 62:315–318

20. Kouhizadeh M, Zhu Q, Sarkis J (2019) Blockchain and the circular economy: potential tensions and critical reflections from practice. Prod Plan Control 31(11–12):950–966. https://doi.org/10.1080/09537287.2019.1695925
21. Krugman P (2010) Building a green economy. New York Times Mag 36
22. Laczniak G (2017) The hidden costs of hidden costs. J Macromark 37(3):324–327
23. Lieder M, Rashid A (2016) Towards circular economy implementation: a comprehensive review in context of manufacturing industry. J Clean Prod 115:36–51. https://doi.org/10.1016/j.jclepro.2015.12.042
24. Markard J, Raven R, Truffer B (2012) Sustainability transitions: an emerging field of research and its prospects. Res Policy 41(6):955–967. https://doi.org/10.1016/j.respol.2012.02.013
25. MacArthur E (2013) Towards the circular economy 1: economic and business rationale for an accelerated transition. J Ind Ecol (Ellen MacArthur Foundation). https://www.ellenmacarthurfoundation.org/es/economia-circular/concepto. Accessed 21 Feb 2021
26. McDonough W, Braungart M (2010) Cradle to cradle: remaking the way we make things. North point press, New York
27. Molina-Moreno V, Núñez-Cacho Utrilla P, Cortés-García FJ, Peña-García A (2018) The use of led technology and biomass to power public lighting in a local context: the case of Baeza (Spain). Energies 2018 11(7):1783. https://doi.org/10.3390/en11071783
28. Molina-Moreno V, Núñez P, Gálvez Sánchez FJ (2019) Transición hacia la economía circular y sostenibilidad de la industria de defensa. Estudio de los casos de Navantia y Airbus Military. Econ Ind 412:149–156
29. Murray A, Skene K, Haynes K (2017) The circular economy: an interdisciplinary exploration of the concept and application in a global context. J Bus Ethics 140(3):369–380. https://doi.org/10.1007/s10551-015-2693-2
30. Nandi S, Sarkis J, Hervani AA, Helms MM (2021) Redesigning supply chains using blockchain-enabled circular economy and COVID-19 experiences. Sustain Prod Consum (Elsevier) 27:10–22. https://doi.org/10.1016/j.spc.2020.10.019
31. Narayan R, Tidström A (2020) Tokenizing coopetition in a blockchain for a transition to circular economy. J Clean Prod 263:121437. https://doi.org/10.1016/j.jclepro.2020.121437
32. Naustdalslid J (2014) Circular economy in China—The environmental dimension of the harmonious society. Int J Sust Dev World 21(4):303–313. https://doi.org/10.1080/13504509.2014.914599
33. Ness D (2008) Sustainable urban infrastructure in China: towards a Factor 10 improvement in resource productivity through integrated infrastructure systems. Int J Sustain Dev World Ecol 15(4):288–301. https://doi.org/10.3843/SusDev.15.4:2a
34. Núñez-Cacho P, Grande FA, Lorenzo D (2015) The effects of coaching in employees and organizational performance: the Spanish case. Intang Cap 11(2):166–189. https://doi.org/10.3926/ic.586
35. Núñez-Cacho P, Leyva-Díaz JC, Sánchez-Molina J, Van der Gun R (2020) Plastics and sustainable purchase decisions in a circular economy: the case of Dutch food industry. PloS one 15(9):e0239949
36. Paolucci B, Hall OA, Axinn NW (1977) Family decision making: an ecosystem approach. John Wiley, New York
37. Pauli GA (2010) The blue economy: 10 years, 100 innovations, 100 million jobs. Paradigm publications
38. Park JY, Chertow MR (2014) Establishing and testing the "reuse potential" indicator for managing wastes as resources. J Environ Manag 137:45–53. https://doi.org/10.1016/j.jenvman.2013.11.053
39. Pearce DW, Turner RK (1990) Economics of natural resources and the environment. JHU Press, Maryland, Pennsylvania
40. Reh L (2013) Process engineering in circular economy. Particuology 11(2):119–133. https://doi.org/10.1016/j.partic.2012.11.001
41. Sedlmeir J, Buhl HU, Fridgen G et al (2020) The energy consumption of blockchain technology: beyond myth. Bus Inf Syst Eng 62(6):599–608. https://doi.org/10.1007/s12599-020-00656-x

42. Seiffert MEB, Loch C (2005) Systemic thinking in environmental management: support for sustainable development. J Clean Prod 13(12):1197–1202. https://doi.org/10.1016/j.jclepro.2004.07.004
43. Stahel WR, Clift R (2016) Stocks and flows in the performance economy. In: Clift R, Druckman A (eds) Taking stock of industrial ecology. Springer, Cham, pp 137–158. https://doi.org/10.1007/978-3-319-20571-7_7
44. Stahel WR, Reday G (1976) The potential for substituting manpower for energy; report to DG V for social affairs. In: Commission of the EC, Brussels (research contract no. 760137 programme of research and Actions on the development of the Labour Market). Brussels
45. Stiehl C, Hirth T (2012) Vom additiven Umweltschutz zur nachhaltigen Produktion. Chem Ing Tec 7(84):963–968. https://doi.org/10.1002/cite.201200008
46. Stoll C, Klaaßen L, Gallersdörfer U (2019) The carbon footprint of bitcoin. Joule 3(7):1647–1661. https://ssrn.com/abstract=3335781. https://doi.org/10.2139/ssrn.3335781
47. Upadhyay A, Mukhuty S, Kumar V, Kazancoglu Y (2021) Blockchain technology and the circular economy: implications for sustainability and social responsibility. J Clean Prod (Elsevier) 126130. https://doi.org/10.1016/j.jclepro.2021.126130
48. Yap NT (2005) Towards a circular economy: progress and challenges. Greener Manag Int 50:11–24
49. Yap NT (2006) Towards a circular economy. Greener Manag Int 50:11–24
50. Yuan Z, Bi J, Moriguichi Y (2006) The circular economy: a new development strategy in China. J Ind Ecol 10(1–2):4–8

The Smart City, Smart Contract, Smart Health Care, Internet of Things (IoT), Opportunities, and Challenges

Ognjen Riđić, Tomislav Jukić, Goran Riđić, Mehmed Ganić, Senad Bušatlić, and Jasenko Karamehić

Abstract The smart city idea is discussed in the literature in multitude of ways. Researcher Komninos specifies the smart cities and surrounding larger regions as environments with a high potential for learning and innovation, utilizing the ingenuity of population and societies with digital infrastructures to function in the physical, uniform, and numerical spaces of cities. The smart city model is defined in the literature in numerous types of ways. As cities develop and expand their services, governance and management are becoming more and more complex. Consequently, cities must adapt to address the economic, social, engineering, and environmental challenges of these transformations. Cities must become smart to face the challenges properly and increase livability and quality of life. This secondary research utilizes detailed literature review of multifaceted sources of information, such as peer-reviewed and quality academic journal articles from renowned databases. With the introduction of blockchain, numerous fields like banking, finance, health care, and supply chain shall experience positive effects. The sustainability of the smart cities can be further enhanced and ensured with the application of the blockchain

O. Riđić (✉) · M. Ganić · S. Bušatlić
International University of Sarajevo (IUS), Hrasnička cesta 15, 71210 City of Ilidža-Sarajevo, BiH, Bosnia and Herzegovina

M. Ganić
e-mail: mganic@ius.edu.ba

S. Bušatlić
e-mail: sbusatlic@ius.edu.ba

T. Jukić
University Josip Juraj Strossmayer, Trg Svetog Trojstva 3, 31000 City of Osijek, Republic of Croatia

G. Riđić
University of Economics for Management (HDWM), Oskar-Meixner-Straße 4-6, 68163 City of Mannheim, Federal Republic of Germany

J. Karamehić
College Center for Business Studies (CEPS), Josipa bana Jelačića 18, 71250 Kiseljak, Bosnia and Herzegovina

S. S. Muthu (ed.), *Blockchain Technologies for Sustainability*,
Environmental Footprints and Eco-design of Products and Processes,
https://doi.org/10.1007/978-981-16-6301-7_7

technology. One important area in which the blockchain represents the important future is real estate and smart cities. Real estate has been going through global transformation. The challenges surfacing in record keeping in scenarios where the same household and/or property is put up for sale to several parties by the means of fraudulently forging the documents or records. The opportunities and challenges must be properly researched, addressed, and weighted out against each other in order to ensure a sustainable future to benefit the most if not all.

Keywords Smart city · Smart contract · Smart health care · Internet of things (IoT) · Blockchain

1 The Smart Contract

The smart contract represents a self-executing digital business procedure. It is numerical agreement which comprises sets of regulations. As such, it operates via disseminated blockchain network. The pallet of regulations is agreed upon by at least two previously well-defined, distinct, and unidentified members. It is implemented by the means of the trigger, either by the particular event and/or demarcated passage of time. The agreement usually entails smooth redistribution of digital type of assets. Furthermore, the proprietorship of digitalized identities of physical assets to enlisted stakeholders is being achieved. This is being affected without involvement of any central third-party execution or human action. The smart agreement implements trusted agreements and corresponding transactions. These transactions are deemed to be (1) transparent, (2) traceable, (3) non-reversible, and (4) dependable. When a smart agreement is put into action, it cannot be meddled with or changed. The major shortcoming of a smart agreement is the restriction of immutable computer protocol to chart real-world contractual resolutions. This is especially true in instances where a disagreement or condition happens, which was not addressed earlier in the smart agreement. In 1994, computer researcher and law academic Nick Szabo pioneered the notion of smart contract. The smart agreement is defined as "a digitalized transaction protocol which executes the contractual terms." Nick Szabo stressed out the enhancement of four basic obligations in legal decisions, listed as follows: (1) privity, (2) discernability, (3) validity, and (4) enforceability. The paramount utilization cases were (1) cost-effective business transactions without a liaison; (2) cost-efficient exchange of synthetic assets and smart property, augmented by self-executing included smart agreements, rescinding the ownership in the scenario when the lease is not paid. Furthermore, the digitalized clauses may be lodged in the smart property to achieve the self-imposition related to the terms of the agreement. The initial application of blockchain aided smart contract was Bitcoin script. Bitcoin script represents a compilation of basic and preordained instructions. These instructions were, on the other hand, significantly constrained in the area of expressiveness. In conclusion, smart agreement achieved eminence with the advent of Ethereum. Ethereum utilized solidity, which represents a Turing complete language broader instruction set, to code the agreements, since the year 2015. Furthermore, every smart agreement is allocated with a distinctive address within a blockchain network. The protocol inside the smart agreement can be seen by each partaking node within the network. In this sense, participating members may opt to engage in the agreement. The smart agreement is typically put into action consequently once a transaction is referred. It can be processed independently via previously defined way including each node in the blockchain network. Smart agreements store up information related to the governmental paperwork, calculates, details, and associations in order to realize the logic implanted in them. Furthermore, the challenges related to the present smart contracts can be summed up as follows: (1) security shortcomings and (2) enforcement in connection to the weaknesses related to scalability [8, 12, 15, 17, 27, 29].

2 The Opportunities in the Application of the Blockchain Technology in Smart Health Care

The growth of novel technologies, examples being Internet of things (IoT) and big data, has affected and enriched the advancement and innovation of health care all over the globe. This growth has enhanced the creation of the smart healthcare system. Smart health care is being exemplified by a medical system including medical cloud data as the foundation. This foundation utilizes (1) the electronic medical record (EMR) and (2) digital health history, in addition to medical IoT. The utilization is carried through by the means of (1) Internet of things, (2) conduction of information, and in addition to the (3) swap over technologies with the aim to construct optimally managed medical and health services. In contemporary times, the smart healthcare industry achieved speedy headway, yet, with the plaguing issues of information and system security. Representing the critical type of technology belonging to the Fourth Industrial Revolution, blockchain retains (1) decentralizing features, (2) secrecy, (3) tamper proofing, and (4) auditability. The unique mixture of blockchain and smart health care may be able to ease the critical pain choke points of traditional smart health care. This may be achieved by its application in (1) data sharing and security, (2) safeguarding of privacy, and (3) enhancement of user-focused smart healthcare systems. Furthermore, the creation of a multiparty medical coalition chain entailing government entities, businesses, and individuals is to stimulate the industrialized modernization of smart health care. Presently, the blockchain has gained the focus of the entire industry. Present research results involve (1) simple blockchain bottom techniques, (2) blockchain key management, (3) long-term authentication evaluation of blockchain signatures, etc. [3–5, 11, 25].

However, the exploration pertaining to the application of blockchain in the smart healthcare arena has not sufficiently matured. Majority of the present studies deal with the amalgamation of blockchain and current information technology to build a new data platform or data system. The examples of these are (1) the creation of an electronic medical system founded on blockchain and (2) the creation of a data privacy safeguard platform centered on blockchain. The studies searching for the role of blockchain in the supportable supply chain of smart health care from the perspective of supply chain are also important to note. For example, the blockchain checking of fake and substandard medications and blockchain supervising of the functioning environment of medical products, in addition to incorporation and enhancement of blockchain relating to a supply chain system are also important to note. Even though the listed literature review depicted a profound research on the application of blockchain in the smart healthcare field, unfortunately, the development system of smart health care under the auspices of blockchain is not yet evident and presently is devoid of the systematic research. In continuation, the involvement of blockchain in smart healthcare industry is rather difficult to be described. From stakeholders' point of view, the search for a multisubject coordinated development system is at the present time rather unique. The contemporary application of blockchain in the area of smart health care predominantly remains at the private chain and shortcoming of exploration at alliance chain level. Since the most dominant firms control the information, in the realm of private blockchains, the information is insufficiently transparent which makes that its application is restricted. Contrasted with the private chain, the public type exudes a larger improvement as far as the informational credibility is concerned. The governmental/public chain ought to involve numerous participating entities. For this and other reasons, this makes it challenging to make sure the privacy and security of participants. Furthermore, the total decentralization impedes the system's design. On the other hand, the so-called alliance blockchain type possesses the distinguishing feature of limited decentralization. As such, it makes it more favorable to the application of blockchain in the realm of smart health care. This is being achieved by inserting limited main body in the application process with the goal of (1) cost reduction, (2) decreased hazards, and (3) increased trust [1, 5].

Smart health care is based on the interaction between patients and medical staff, medical institutions, and medical devices by creating a health archives regional medical data platform. It utilizes the most sophisticated IoT technology. By doing so it ensures that the medical industry increasingly obtains the necessary information. Sharing of medical data is an essential step to make the medical system more intelligent and increase the quality of medical service. Furthermore, the distribution of patients' data between organizations has not been completely achieved. In this sense, the blockchain represents a potent mean to resolve this issue. Blockchain represents various distributed data systems including multiple independent nodes. As such, it includes (1) decentralization, (2) time stamps, (3) collective maintenance, (4) programmability, and (5) tamper proofing. Blockchain involves comparatively few medical treatment-related applications. The contemporary research primarily places attention on the combination of blockchain with a particular information technology. The goal of this to create a single application podium. Utilization of blockchain technology to construct a medical transaction pane verification system in addition to utilization of the etheric blockchain to build a medical information sharing platform MedRec associates the blockchain with big data. All of this is achieved by utilizing blockchain technology in connection to the OPAL/Enigma encryption stage to build a secure environment for medical information loading and analysis. Consequently, the application of blockchain technology in the entire intelligent medical industry is being plagued by the missing systematic research [5].

3 Opportunities in Utilizing Smart Contracts in Smart Health Care

Following the introduction of the General Data Protection Regulation (GDPR) in May of 2018, the security, privacy, transparency, and consent for patient-owned medical data became the focus of the healthcare institutions' anxieties. It is important to note that the overt patients' consent related to the processing of health data and the transparency as to how information shall be collected, in which way it will be collected and patients' rights to full access to their health information have significantly impacted healthcare information systems (HIS). On top of the data generated by healthcare institutions, the patients tend to be increasingly active in managing their diseases by gathering health data utilizing mobile devices in addition to the sensors. Disseminating patients' self-collected information with medical systems has a positive effect on disease management. Blockchain technology has been gaining broad exposure in health care. As such, it promised significant advancements, examples of which are smart healthcare management and patient empowerment. There are also smart contracts realized utilizing blockchains, and they are commonly known as Blockchain 2.0. They represent types of protocols which allow the authentication and execution of legal agreements involving two or more legal parties by making them irrevocable. Interest in smart contracts has been gaining attention ever since the introduction of Ethereum. Ethereum, publicly released in 2015, represents the original blockchain-based solution incorporating smart contracts. Smart agreement permits patients to manage (1) access to their health archives and (2) safe exchange of data, in addition to safeguarding the privacy of the discussed exchanges. Potential utilization of smart contracts in health care, the intentions, and limitations, aiming at information sharing, and deliberation as to why there is still lack of practical applications in a real situations are important areas for the future research [6, 8, 19, 20, 24] (Table 1).

The enhancement of blockchain relating to the smart healthcare system is largely mirrored within nine distinctive aspects, being listed as follows: 1. design at the highest level, 2. physician management, 3. medical records management, 4. treatment optimization, 5. community building, 6. cost savings, 7. internal and external regulation, 8. medical insurance, and 9. governance of the environment. These nine distinctive facets were found to possess varying levels of importance. The design at the top level, the medical records management, and the physician management represent the underlying causes of the system construction. In continuation, the particular application of blockchain in the realm of smart health care is primarily focused on the area of the intelligent contract. It depends on the management of medical records,

Table 1 Areas of the application of the blockchain technology in smart health care

Aspects	Criteria	Explanation
External regulation (Al)	Medical supply chain regulation	The regulation department supervises the drug and equipment supply chain through the blockchain
	Medical process regulation	Regulators monitor medical processes through the blockchain
	Regulation of clinical waste treatment process	The medical waste disposal process is monitored by the regulatory authority through the blockchain
Medical record management (A2)	Electronic medical record	Using blockchain technology to share electronic medical records: customers have absolute ownership
Treatment optimization (A3)	Targeted therapy	Using blockchain technology to share patient data; drug discovery agencies are licensed to target drugs
	Telemedicine	The use of blockchain and Internet of Things and other technologies for telemedicine
Doctor management (A4)	Identification	Using blockchain technology to build a verification platform, the doctor's identity, and certificate for periodic verification
	Personnel Screening	Using blockchain to collect and update the diagnosis results and correct rate of doctors in real time and to screen the excellent doctors
	Customer choice	Using blockchain to store the doctors identity information and treatment information; it is also helpful for customers to choose the right doctors
Medical insurance (A5)	Maintenance of interests	Insurance companies, customers, and medical
	Resource control	Institutions through the blockchain intelligent contract insurance transactions to protect the interests of customers
	Enhancing risk management capacity	Reduce the hospital's unpaid bills

(continued)

Table 1 (continued)

Aspects	Criteria	Explanation
Internal regulation (A6)	Regulation of drugs and medical equipment	Using the blockchain to set up a tracking system; medical institutions can monitor the flow of drugs and medical equipment in real time
Cost saving (A7)	Cost saving in doctor-patient communication	Using blockchain to share electronic medical records and saving communication cost
	Cost saving of drugs research and development	Data management system based on blockchain intelligent contract technology to reduce the management cost of multisite clinical trials
Top-level design (A8)	Industry standards	Industry associations establish industry standards
	Credit rating system	All nodes are rated by the industry consortia
	Reward and punishment system	The node that obeys a rule is given reward, and the node that does not obey a rule is given punishment
Community structure (A9)	Health management	Using blockchain to collect customer health data and create community health programs
	Family sickbed	Using blockchain and Internet of things technologies for remote monitoring and treatment at home
	Hierarchical diagnosis and treatment	Community doctors perform primary care, analyzing the patient's physical condition and deciding whether to go to a hospital
Environmental management (A10)	Clinical waste treatment	Through blockchain and Internet of Things technology to upload medical waste treatment data and improve treatment efficiency

Source (Du, Chen, Ma, & Zhang, 2021). [5]

controlled by the system, and adjusting the application is paramount in upgrading of the system. The regulation and medical insurance, connected with the governance of the environment, represent a defending role system's advancement. All of them efficiently protect stakeholders' interests. The efficiency of the internal and external supervision represents a key element in the health of the system. In conclusion, the blockchain run intelligent medical application system ought to be built centered around three distinctive layers: (1) the stakeholder layer, (2) information layer, and (3) transaction layer [5, 7].

4 Smart Contract Applications

Among numerous thought-provoking applications utilizing the concept of blockchain, the most interesting applications tend to be smart contracts which encompass a programmable code which is being triggered into execution mode, only in instances when particular circumstances in the blockchain network are being achieved. "Smart contract" was first pioneered by researcher Szabo [29]. From the start, this became an extraordinary concept to methodically execute agreements among the linked legal parties. A smart agreement is utilized to produce a contract between two or more parties that implement a predetermined course of action. Within the blockchain, a smart agreement is performed in a decentralized manner. The agreements are being executed by the blockchain only in instances when particular conditions are being satisfied. To reach the approval of both parties, the contract includes comprehensive explanation of legal terminology and agreements. This agreement is significantly more complex and necessitates legal authority as a mediator to assist the parties to reach common agreement via signature of both parties. The blockchain promises to build these contracts in a more convenient and secure digital way [1]. The subject matter of the contract may be authenticated utilizing cryptographically hashed keys deposited in distributed ledger shared by nodes in the blockchain network. The sizeable description of terms may be kept with two involved legal parties. The authenticity of those before-mentioned terms could be validated by other members in the network by calculating hash of terms and comparing it with cryptographically hashed key stored in the ledger. Secure contracts solutions can be provided by further validation of cryptographically hashed key [12]. In its research study, Luu et al. [15] projected that secured contracts in the blockchain network could additionally remove the need for public notaries in future. Smart contracts may be utilized to deliver services such as asset trade contracts, insurance contracts, legal contracts, and registry contracts. Researchers pioneered an automated signing protocol among two authorities by employing decentralized network as a way for delivering time-stamped types of utilities. A smart contract may be utilized in various areas of businesses for distinct requirements. Kishigami et al. [14] introduced a digital content distribution system built on the blockchain. They surveyed 100 participants containing digital content stakeholders and creators. The results showed that decentralized mechanism tends to be the most influential factor for digital rights management approach [14, 22].

Numerous kinds of transactions relating to smart cities may be recorded in a blockchain. By utilizing smart contracts, convoluted legal procedures may be performed, and data exchange can be executed automatically. By using smart contracts and decentralized applications, the blockchain provides a high level of

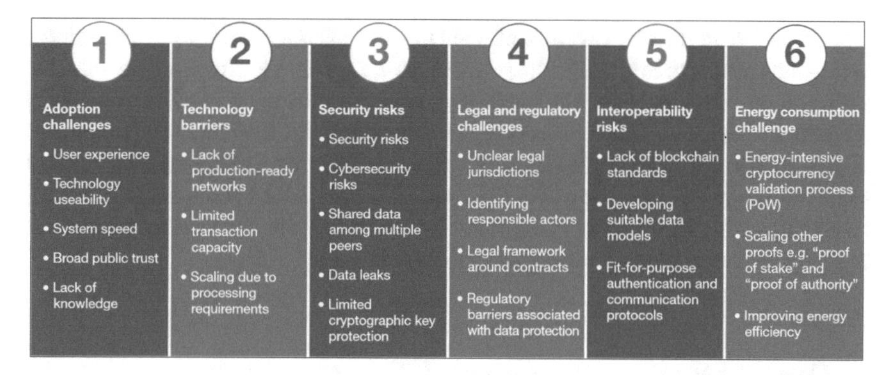

Fig. 1 High-level summary of various blockchain risks and challenges. *Source* Manushaqa et al. [18]

autonomy in processing of the smart transactions for the duration of the operational process in the smart city. Blockchain may also provide features, such as flawless privacy, authentication, effortless deployment, and maintenance and security [13, 17].

5 Blockchain Implementation Challenges in Smart Cities

Figure 1 depicts-high level summary of blockchain (implementation) risks and challenges.

These could be summarized as the adaptation challenges (i.e., user experience or the lack of knowledge), technology barriers (e.g., limited transaction capacity), security risks (e.g., data leaks and limited cryptographic key protection), legal and regulatory challenges (i.e., unclear legal jurisdictions), interoperability risks (e.g., lack of blockchain standards), and energy consumption challenges [18].

6 Smart City, Blockchain, and Internet of Things (IoT) Opportunities and Challenges

A smart city strives to resolve challenges which are instigated by population growth in urban areas. These areas include 1. physical security, 2. residual management, and 3. transportation systems, utilizing technology to interconnect governmental organizations. They are permitting citizens' right to use the multiple services, by supporting a path to manage various public resources in an effective and transparent manner. The Internet of things (IoT) technology presents low-cost and efficient resolutions for the development of smart cities. IoT achieves this due to its robust applicability in a limitless pallet of scenarios. The intelligent IoT gadgets are able to constantly work together in different sectors of a smart city scenario. They are exchanging a never-ending flow of information to provide quality services to citizens as its supreme goal. IoT devices are deemed as intelligent, since they are capable of autonomously communicating, with little or even without any human involvement. The promotion of IoT solutions involves numerous security risks, having in mind the shortage of supervision and the restriction of computational resources in IoT devices. Numerous research articles have focused their attention on uncovering the weaknesses of IoT ecosystems, while recommending efficient resolutions [2, 26].

7 Security Issues with Internet of Things (IoT) Devices in Context of Smart Cities

A smart city includes the implementation of various services presented in situations, such as 1. smart transportation systems, 2. smart hospitals, and 3. smart airports. Every case is reinforced by IoT devices in a presented sequence ($D = \{D1, \ldots$ DnD$\}$). They possess the capability of swapping data utilizing either local or remote devices stationed on the Internet. IoT gadgets may be utilized for various aims. They contribute to automated procedures due to their ability to capture and process information in real-time manner. Furthermore, in particular instances, it is required to share information processed by IoT devices (e.g., sensed values) utilizing a software or hardware component utilized in different scenarios. They could use this type of information to start a process. In this scenario, IoT gadgets present a week point vulnerable to vicious attack. This can be explained by their robust presence in smart cities, in connection to the decreased capability to act against a cyber-attack. Keeping in mind the effects which a cyber-attack could present to IoT gadgets and smart city services, we may clearly infer the notion that before-mentioned gadgets and their associated infrastructure ought to be protected. Consequently, their security shall guarantee that any potential attacker striving to attempt an unauthorized action/attack related to IoT devices shall be precisely traced back [2].

As previously discussed, the blockchain technology contains multiple security properties, being non-repudiation and traceability among the most notable ones. Thanks to these properties, all registered information in BlockSIEM is permanently available and can be consumed in future by any SIEM enabled as a miner (Mi). Besides the security events, each block also contains metadata, such as the ID of the IoT sentinel that created the security events, the creation date of the security events, and any other information that may be useful in the auditing process of a security incident. The inputs related to the IoT sentinel ID discovering the threat and reports of the security events combined with other information regarding the security event report may be consulted from the blockchain in a secure and trustworthy way, since such information is protected against modification, making it useful for forensic purposes [2].

8 Challenges in Connecting Citizens and Smart Cities

The population has been increasing rapidly around the world, and, along with this growth, new strategies are being planned to provide a better life for humankind. New initiatives and novel technologies are emerging in various manners, promoting a new wave of innovation within cities' services. However, even with continuous advance of cities and digital technologies, different challenges to engage citizens in social decisions are still open and need to be constantly improved. The concept of "smart city" (SC) was pioneered in the 1990s. It opened new viewpoints as to how new technology may affect the cities. Researchers Dameri and Cocchia pioneered the synopsis on the history of the classification "smart". It points out to the pioneering Gibson et al. study entitled The Technopolis Phenomenon: Smart Cities, Fast Systems, Global Networks. Smart cities combined with ICT and urban evolution. Smart city concept is two decades old. It has been developing in a steady manner. The most important purpose of smart cities is in facilitation of the 1. planning and 2. construction of the intelligent services in the cities. On the other hand, this concept is also burdened with significant amount of skepsis due to its futuristic properties so that it is frequently being labeled the "smart utopia" [4, 10, 19–21, 28, 30].

9 Smart City, Blockchain, and Internet of Things (IoT) Opportunities

In order to address the security challenges posed by an IoT ecosystem, BlockSIEM proposes a blockchain-constructed and disseminated security information and event management (SIEM) framework to safely detect, store, and analyze IoT security events. All of this is being achieved while keeping their integrity and non-repudiation elements. In this sense, due to the blockchain technology, BlockSIEM showcases numerous security properties. Non-repudiation and traceability represent the most important features. Non-repudiation implies that the blocks are immutable. In this sense, an entity is not able to refuse that it generated a transaction request. The traceability of the information logged in the blocks allows to identify the trail of a specific security event. Furthermore, the BlockSIEM may be employed in various utilization scenarios. All of this ensures the 1. trustworthiness, 2. resilience, 3. Scalability, and 4. auditability of the system. BSIEM-IoT, a blockchain-founded SIEM resolution which was deemed appropriate for IoT environments, was proposed, generating security events thanks to the effort of IoT devices called sentinels and reporting these through blockchain to be read and analyzed by SIEM miners having internal and external threat intelligence capabilities. The main contributions of it were the initial proposal of BSIEM-IoT as a blockchain-based SIEM, enclosing a smart contract used to handle blocks of security events to prevent and detect attacks over IoT scenarios. BSIEM-IoT used external and internal threat intelligence with the aim of detecting distributed attacks [2, 16, 23, 25].

Numerous suggestions were proposed in the last years with the goal of safeguarding IoT ecosystems. One type of security architectures founded on the utilization of security events relies on a multi-relation between the attack-associated rudiments. There are security events categories, providing information about the impact of an attack over a given IoT gadget. They are being combined with weaknesses, to explicate the root causes of the attack, in addition to the explanation of attack surfaces with the aim to provide the information through which avenues the attack was undertaken. On the other hand, numerous blockchain-based solutions have been proposed with the aim of leveraging the security attributes of the distributed ledgers within IoT scenarios. An IoT security framework for a smart home scenario is based on the reliance on a hierarchical structure that coordinates methods over the blockchain network to keep the security and privacy advantages presented by this technology. Such a hierarchical structure is more appropriate for the particular requirements of IoT. The tasks on the network are being carried out in a different and adjusted manner than a usual kind of blockchain, example being the Bitcoin. The framework proposes to manage the network, and the belonging devices with the methods store, access, monitor, genesis, and remove. This framework was exemplified in the context of a smart home, but the application is opened to be used to other IoT contexts [2, 9, 24].

10 Conclusion

Fast urban growth brings to the table novel infrastructure issues for public administrations. As cities develop and expand their services, governance and management tend to be increasingly complex. Consequently, cities must adapt to address the economic, social, engineering, and environmental challenges associated with these alterations. Cities must become smart to confront the challenges in a proper manner. Although it was originally built as a digital platform for cryptocurrencies and financial transactions, blockchain promises potential applications in various areas, involving city management. A smart city depends on the components that permit data collection and storage, backed by the specialized hardware infrastructure. An important aspect to consider is ensuring the interoperability of new applications with existing ones, so it is reused by the existing infrastructure, where possible. A smart city uses digital technologies to improve the quality and performance of urban services for citizens. Citizens' access to smart city applications must be guaranteed in an interactive way, so that they can access and transmit, as the case may be, real-time information. City officials and local administrations of a smart city based on a blockchain would benefit from a completely new way of connecting with citizens and visitors.

This secondary research utilizes detailed literature review of multifaceted sources of information, such as peer-reviewed and quality academic journal articles from renown databases.

Studies point out that the increased rate of urbanization is putting pressure on the infrastructures. In order to support the growing population in the big cities, a new type of smart infrastructure is necessary. This type of infrastructure shall be capable of supporting the efficiency and effectiveness of services and quality of life. Blockchain is a promising solution for a wide range of challenges faced by a smart city, but the implementation depends on the city administration and the needs of the community. A blockchain-based smart city infrastructure has the advantages of increased efficiency due to the automated interactions with its citizens, optimized distribution of resources, and fraud reduction.

Blockchain and smart cities concepts are destined to shape our future in different ways, but they could shape it by working together. The future cities that we will live in will be more interconnected than ever. IoT devices and sensors will help transmit data, create efficiencies, and adapt our living habits to our environment. Incorporating blockchain into the development of smart cities will make it possible to have a cross-cutting platform that connects the cities' different services, adding greater transparency and security to all services and processes.

With the introduction of blockchain, numerous fields like banking, finance, health care, and supply chain shall experience positive effects. The sustainability of the smart cities can be further enhanced and ensured with the application of the blockchain technology.

References

1. Bigi G, Bracciali A, Meacci G, Tuosto E (2015) Validation of decentralized smart contracts through game theory and formal methods. Program Lang Appl Biol Secur 9465:142–161
2. Botello JV, Mesa AP, Rodríguez FA, Díaz-López D, Nespoli P, Mármol FG (2020) BlockSIEM: protecting smart city services through a blockchain-based and distributed SIEM. Sensors (Basel, Switzerland) 20(16). https://doi.org/10.3390/s20164636
3. Ćirić Z, Ivanišević S (2019) Implementation of blockchain technology in the smart city. In: Conference: 5th international scientific conference on knowledge based sustainable development—ERAZ 2019At. Budapest, Hungary
4. Davidson S, De Filippi P, Potts J (2016) Economics of blockchain. https://hal.archives-ouvertes.fr/hal-01382002/document/. Accessed 20 Jan 2019
5. Du X, Chen B, Ma M, Zhang Y (2021) Research on the application of blockchain in smart healthcare: constructing a hierarchical framework. J Healthc Eng 2021:6698122. https://doi.org/10.1155/2021/6698122
6. Durneva P, Cousins K, Chen M (2020) The current state of research, challenges, and future research directions of blockchain technology in patient care: systematic review. J Med Internet Res 22(7):e18619. https://doi.org/10.2196/18619
7. Ehrenberg AJ, King JL (2020) Blockchain in context. Inf Syst Front 22(1):29–35. https://doi.org/10.1007/s10796-019-09946-6
8. Giordanengo A (2019) Possible usages of smart contracts (blockchain) in healthcare and why no one is using them. Stud Health Technol Inform 264:596–600. https://doi.org/10.3233/SHTI190292
9. Grover P, Kar AK, Janssen M (2019) Diffusion of blockchain technology: insigshts from academic literature and social media analytics. J Enterp Inf Manag 32(5):735–757. https://doi.org/10.1108/JEIM-06-2018-0132
10. Hoxha V, Sadiku S (2019) Study of factors influencing the decision to adopt the blockchain technology in real estate transactions in Kosovo. Prop Manag 37(5):684–700. https://doi.org/10.1108/PM-01-2019-0002

11. Hussien HM, Yasin SM, Udzir SNI, Zaidan AA, Zaidan BB (2019) A systematic review for enabling of develop a blockchain technology in healthcare application: taxonomy, substantially analysis, motivations, challenges, recommendations and future direction. J Med Syst 43(10):1–35. https://doi.org/10.1007/s10916-019-1445-8
12. Idelberger F, Governatori G, Riveret R, Sartor G (2016) Evaluation of logic-based smart contracts for blockchain systems. In: international symposium on rules and rule markup languages for the semantic web. Springer, Cham, pp 167–183
13. Khan P, Byun Y-C, Park N (2020) A data verification system for CCTV surveillance cameras using blockchain technology in smart cities. Electronics 9(3):484 (MDPI AG). https://doi.org/10.3390/electronics9030484
14. Kishigami J, Fujimura S, Watanabe H, Nakadaira A, Akutsu A (2015) The blockchain-based digital content distribution system. In: IEEE fifth international conference on big data and cloud computing, Dalian, pp 187–190
15. Luu L, Chu DH, Olickel H, Saxena P, Hobor A (2016) Making smart contracts smarter. In: Proceedings of the 2016 ACM SIGSAC conference on computer and communications security association for computing machinery, Vienna, pp 254–269
16. Mackey TK, Kuo T-T, Gummadi B, Clauson KA, Church G, Grishin D, Obbad K, Barkovich R, Palombini M (2019) "Fit-for-purpose?"—Challenges and opportunities for applications of blockchain technology in the future of healthcare. BMC Med 17(1):68. https://doi.org/10.1186/s12916-019-1296-7
17. Majeed U, Khan LU, Yaqoob I, Kazmi SM, Salah K, Hong CS (2021) Blockchain for IoT-based smart cities: recent advances, requirements, and future challenges J Netw Comput Appl 181 https://doi.org/10.1016/j.jnca.2021.103007
18. Manushaqa L, Grant M, Baliakas P, Holotescu T, Amellal J (2019) Blockchain implementation in smart cities: With a Deep Dive Use Case: Smart Dubai. Module DFIN524 - Blockchain Applications.
19. Oliveira TA, Oliver M, Ramalhinho H (2020) Challenges for connecting citizens and smart cities: ICT, E-governance and blockchain. Sustainability 12(7):2926 (MDPI AG). https://doi.org/10.3390/su12072926
20. Patrício LD, Ferreira JJ (2020) Blockchain security research: theorizing through bibliographic-coupling analysis. J Adv Manag Res 18(1):1–35. https://doi.org/10.1108/JAMR-04-2020-0051
21. Rodríguez B, Bolívar R, Scholl HJ (2019) Mapping potential impact areas of blockchain use in the public sector. Inf Polity: Int J Govern Democr Inf Age 24(4):359–378. https://doi.org/10.3233/IP-190184
22. Rotuna C, Gheorgita A, Zamfirou A, Smada DM (2019) Smart city ecosystem using blockchain technology. Inf Econ 23(4):41–50. https://doi.org/10.12948/issn14531305/23.4.2019.04
23. Saez MIG (2020) Blockchain-enabled platforms: challenges and recommendations. Int J Interact Multimed Artif Intell 6(3):73. https://link.gale.com/apps/doc/A638116052/AONE?u=uphoenix&sid=AONE&xid=efccb900
24. Sakho S, Jianbiao Z, Essaf F, Mbyamm Kiki MJ (2019) Blockchain: perspectives and issues. J Intell Fuzzy Syst 37(6):8029–8052. https://doi.org/10.3233/JIFS-190449
25. Sanka AI, Irfan M, Huang I, Cheung RCC (2021) A survey of breakthrough in blockchain technology: adoptions, applications, challenges, and future research. Comput Commun 169:179–201. https://doi.org/10.1016/j.comcom.2020.12.028
26. Singh P, Nayyar A, Kaur A, Ghosh U (2020) Blockchain and fog based architecture for internet of everything in smart cities. Future Internet 12(4):61 (MDPI AG). https://doi.org/10.3390/fi12040061
27. Stoica M, Mircea M, Ghilic-Micu B (2020) Using blockchain technology in Smart University. E-Learn Softw Educ 3:134–141. https://doi.org/10.12753/2066-026X-20-187

28. Swan M (2015) Blockchain: blueprint for a new economy. O'Reilly Media, Sebastopol, CA
29. Szabo N (1997) The idea of smart contracts. Nick Szabo's Papers and Concise Tutorials, 6
30. Veuger J (2018) Trust in a viable real estate economy with disruption and blockchain. Facilities 36(1/2):103–120. https://doi.org/10.1108/F-11-2017-0106

Applications of Blockchain Technology for a Circular Economy with Focus on Singapore

Sareh Rotabi and Omar Ali

Abstract As a new economic paradigm, circular economy (CE) has recently gained attention. CE aims to detect, recover, sustain or increase value by transitioning from a 'take, make and waste' approach of production system to an ecosystem economy, thus restraining the environmental effect and waste of resources throughout the product economy, providing profitability and flexibility and creating value of what is considered a waste. Cities' role in driving CE forward has been observed as being increasingly important to reach Sustainable Development. For this reason, there is an urgency to understand what a circular city is and how it might be composed. On the other side, blockchain is a method of recording information in a way that makes it difficult or impossible to change, hack or cheat the system. This chapter initially explores the definition and characteristics of blockchain in detail. Then, several application scenarios of blockchain technology for a CE with focus on Singapore while transitioning to a CE are put forward. Finally, some of the challenges including energy consumption, lack of adequate skill set, the cost associated with the advanced technology and the need for raising awareness of circular business models, that still exist when applying the current blockchain technology to the CE, are discussed and analysed.

Keywords Blockchain · Innovation · Circular economy · Resources · Sustainability · Singapore

---

S. Rotabi (✉)
Department of Quality Assurance and Institutional Advancement, American University of the Middle East (AUM), 54200 Egaila, Kuwait
e-mail: Sareh-Rotabi@aum.edu.kw

O. Ali
Department of MIS, Business and Administration College, American University of the Middle East (AUM), 54200 Egaila, Kuwait
e-mail: Omar.Ali@aum.edu.kw

S. S. Muthu (ed.), *Blockchain Technologies for Sustainability*,
Environmental Footprints and Eco-design of Products and Processes,
https://doi.org/10.1007/978-981-16-6301-7_8

1 Blockchain

Blockchain is regularly discussed as the ground-breaking innovation and an indication of a new economic period [9]. Technology supporters discuss that blockchain creates the base for trustworthy distributed record systems that could be essential to how personal and inter-organizational relationships are planned [7]. It is likely to have special effect on relationship, which is mainly caused by the ability of the technology to allow unaffectedly trust-free social and economic transactions that are imposed via a confirmed and transparent record and value exchange mechanisms without the support of a central authority [8].

Blockchain is likely to shake a multiple of industries and businesses as a foundation technology [53], where it assists as decentralized infrastructure, whether it is for implementation of a certain jobs (e.g. for financing, payments, investments, notary services) [1] or as practical support, is capable of helping markets [68, 84, 85, 93]. The demand to set blockchain in such type of circumstances not only outlines the point that the technology reduces the use of intermediaries, in fact it is usually anticipated to develop the security of particular infrastructures and records, to expedite transparency and ability to track transactions, which altogether result in decreasing of the operational costs [52].

Great hopes concerning the capability of blockchain for the public and businesses activated a range of design-oriented research in the information system (IS) field (e.g. [52, 84]). Simultaneously, IS researches condemn the results of these design studies. Actually, the outcomes of the IT articles are usually built upon a group of assumptions related to sociotechnical subjects, namely through disregarding the governance and coordination cost, neglecting the privacy issues along with the risk factors concerning the efficiency and efficacy of a clever agreement. Accordingly, the result of these articles continues to be subjected for speculation [9]. Numerous researchers were approached lately for further acute viewpoints on blockchain in the IS researches (e.g. [7, 68, 93]), causing a discussion related to how and where blockchain is successfully valid and where it could offer a special social effect [93].

In the meantime, the research of IS related to the blockchain came from hype to certainty through confessing that there is a shortage of systematic, technology-driven methods of understanding the possibility of technology [39]. To overcome this shortage of the knowledge, it needs an inclusive indulgence and evaluation of the special effects of the particular parts of the blockchain and an analysis of common sides of the application [93]. Yet, the empirical studies that are based on theory outside the blockchain's primary use situation as the practical support of the cryptocurrency Bitcoin is limited [25, 93]. This could be due to the scholars' shortage of knowledge on how to group data for effective quantitative exploration or theoretical fuzz of the blockchain occurrence [47, 93].

The ability to collect information for quantitative research is low, which is mainly due to the fact that companies and researches in general tend to find difficulty in recognizing sufficient use cases for blockchain. This reduces the availability of data for scholars [39, 64].

It is essential to conduct empirical studies in the area of the role and impacts of the blockchain for society and business as it connects theory and practice together and aids in recognizing the use cases where the technology influences efficacy and the overall efficiency level increases [86]. It is important to conduct such type of researches taking to account that the practical application of blockchain is yet rare, and companies' investments in the field of technology during the period of 2012–2017 were about $1.2 billion. This includes large organizations, namely Citi, JP Morgan, NYSE, Visa, Wells Fargo, USAA, Wal-Mart and MasterCard [64].

This chapter initially explores the definition and characteristics of blockchain in detail. Then, several application scenarios of blockchain technology for a CE with focus on Singapore while transitioning to a CE are put forward. Finally, some of the challenges including energy consumption, lack of adequate skill set, the cost associated with the advanced technology and the need for raising awareness of circular business models, that still exist when applying the current blockchain technology to the CE, are discussed and analysed.

1.1 Blockchain Overview

There exist different definitions for the blockchain, starting from application specific to the extremely technical. For instance, the world's biggest cryptocurrency exchange, Coinbase, describes blockchain as follows: 'a distributed, public ledger that contains the history of every bitcoin transaction' [20]. From an application-specific definition, this description does not reflect the point that blockchain could be reprocessed for other cryptocurrencies and industry applications autonomously. According to the Oxford English Dictionary, the definition of the blockchain is slightly broader and described as follows: 'a digital ledger in which transactions made in bitcoin or another cryptocurrency are recorded chronologically and publicly' (Oxford [87]). According to Sulan [108], this definition is also limited as blockchain technology could be used independently from cryptocurrencies and Bitcoins. The above-mentioned definitions focus on the role of a blockchain as a digital ledger, yet this field is still rising and growing fast. The usage of the ledger is one of the features of the blockchain but not its core. This feature mainly affects blockchain application that emphasizes on the interchange of value management in case of virtual assets.

Another wider definition by Webopedia defines blockchain as follows: 'a type of data structure that enables identifying and tracking transactions digitally and sharing this information across a distributed network of computers. The distributed ledger technology offered by blockchain provides a transparent and secure means for tracking the ownership and transfer of assets' [106].

Even though this description prospers the main characteristic of the blockchain, it focuses on the distribution as the core feature of the blockchain computing, yet lacks the acknowledgement that blockchains are not only a distributed technology, moreover, a decentralized one [109] (Wright and De Filippi 2015).

According to Sulan [108], the main distinction here is that the distributed system divides the task among contributors in an ideal manner while a blockchain demands from every contributor to sustain a full bulge of the system and impose its regulations autonomously. In a system which bulges function on domestic information to achieve goals instead of the result of the central assembling impact, this decentralization guarantees that dragging the plug is almost not possible. In order for the network to operate, only one node is needed to continue functioning. Clearly, there is a demand for a strong and brief definition of blockchain. Taking into account previous definitions, blockchain can be defined as a 'a decentralized database containing sequential, cryptographically linked blocks of digitally signed asset transactions, governed by a consensus model.'

Sulan [108] believes that this definition aims to focus on the core elements for blockchain technology where it is a peer-to-peer network database ruled by set of regulations. In addition, blockchains embody a moving away from a conventional trust agent and shift towards transparency. Blockchain allows applications from a wide range of industries to benefit from sharing, following and assessing digital assets.

1.2 Characteristics of Blockchain

According to Sulan [108], even though blockchain was initially connected with Bitcoin, it can also be used autonomously in a diversity of markets and different use cases, namely health and insurance industry. In fact, blockchain can be implemented on any sector where assets are managed and transactions take place. It actually offers a safe chain of custody for the physical assets and so as the digital ones via its practical features that smooth the transactions through smart contracts, security, consensus and trust. More about each of these aspects are explained further below:

Transactions and Smart Contracts—Transaction occurs via exchange of assets which are managed through the company services' guidelines. These guidelines are reflected via scripting languages (e.g. Bitcoin's forth) and are operated for progressive transactions that need to be implemented, where they also build the foundation for smart contract. In order to manage a transaction, a smart contract, which is a group of logic guidelines in the form of a coded script, needs to be inserted. The contract is implemented independently and is used to manage the transaction [15]. In this method, contract performs like smart software agents [104]. After a smart contract is implemented in the blockchain, it turns into an independent agent that is unable to be tampered. Following this stage, an application then recites the codes when executing a transaction, perform and process the outcomes. The nature of the smart contract is not limited to an application-specific code. It is also used to classify the condition and terms of a contract into the workflow of the transaction. Ethereum (the second-largest cryptocurrency after Bitcoin) is an alt-coin technology that has been developed to support smart contracts.

Consensus and Trust—Blockchain is trusted by agreement as all related parties should have the same copies of the blockchain, yet each member still have to validate it. The core strength of the blockchain is in its decentralization aspect taking into account its database is owned by all members. To make sure that each copy is reliable, an agreed algorithm is needed. The agreed algorithm permits the public to confirm that every single added block is authentic. It also stops the attackers from bargaining and splitting the chain [21]. Another approach recommended was to use proof-of-work method where a hard cryptographic puzzle needs to be resolved by miners [79]. Miners consume resources and are acknowledged for their work through several motivations. There are also other models, namely proof-of-burn, proof-of-elapsed-time, proof-of-stake and proof-of-capacity which have been recommended in the literature to handle the flaws of the initial proof-of-work model by trying to equalize objectivity and source spending [57]. According to Sulan [108], blockchain by nature is suspicious. It has been created to reduce the necessity for any object to enter transactions. It builds a reliable model according to a group agreement, as the network confirms the transactions and approved their adding to the chain. There is no intermediary, hence the idea of trust continues to be inherent as every record in the blockchain is confirmed by the public that carries serval copies of the blockchain. When trust mediators are eliminated from transactions, blockchain is capable of interrupting several main businesses. The customary transaction models count on dominant expert to perform in the clearinghouse character. Trust is given to the dominant expert with a probability that it will continue to be truthful during clearing transactions and validation. The occurrences of records are included with the dominant expert. In case the dominant expert is cooperated deliberately, such as hacking, then the interlocutor has the ability to inflict a widespread chaos into the system. The blockchain model reduces the dominant expert through distributing copies of the records to all related parties (Fig. 1). Every single contributor keeps their own case of the blockchain. They transmit the variations through creating fresh blocks and ask for confirmation as per the agreed model rules. As soon as it is confirmed, the block would be added to everyone's chain. The method is likely more

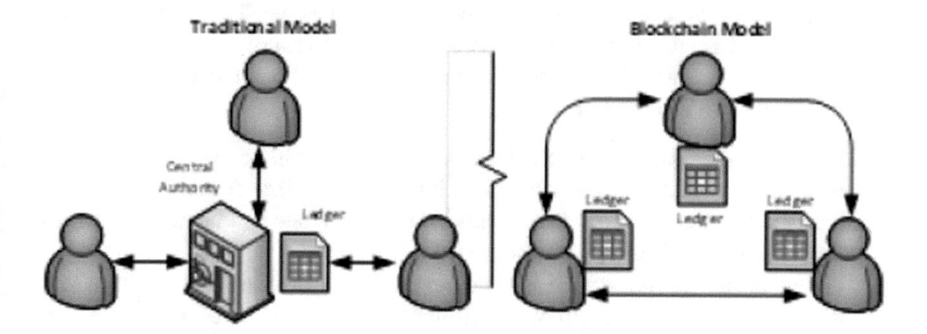

Fig. 1 Traditional versus blockchain transaction model (Sulan et al. 2018)

secure than the customary model, and the intermediary agent is not needed anymore, which raises an interruption to the current situation.

A blockchain counts on arithmetic through cryptography to build autonomous confident for every single transaction and on computationally exclusive agreed model to substitute dominant experts. Many of the latest transactions are well ordered into the block. Afterwards, the block is cryptographically connected to group of blocks, called blockchain, and is confirmed via an agreed model that includes major computing assets. Since the blockchain is a file with open-access feature and simulated on several full nodes of the system, there is no object that can control the list of transaction. As each block is shredded and implanted into the chain, it is undeniable and acts as the last record of the previous transactions. An object is not able to modify the chain unless it changes all the other blocks that are after it, and this is a hard and costly computational effort. Hence, this safeguards the blockchain and builds a truthful autonomous of a dominant expert (Sulan [108].

Public and Private Blockchains—Blockchain can be categorized as private, public or hybrid, which relies on their use [14, 78]:

- Private: These blockchains refer to the rights to govern who can read from and inscribe to the blockchain. Agreed algorithms and mining are generally not needed as the single object has the right and governs the creation of the block.
- Public: These types of the blockchains do not have an owner and are noticeable by everyone. Their agreed method is exposed to everyone to contribute in, and they are completely decentralized. An example of a public blockchain is Bitcoin.
- Hybrid: These blockchains are also called as consortium and are available at the public but only to a private group. The agreed method is governed by certain servers using a group of rules that are accepted by all related parties. And, the copies of the blockchain are solely circulated between allowed contributors, hence the system is not fully decentralized.

Even though, the public blockchain circulates itself in a decentralize peer-to-peer mode, this is not applicable for a private blockchain. Private blockchains are the ones used by initiatives to record the transactions of the asset inside a partial user base. As for the hybrid blockchains, they can be envisioned as small scale of the public blockchains and are decentralized mainly to a restricted group.

According to Sulan [108], in a brief, we are able to reinforce the main features of the blockchain into four main features:

1. Decentralized: A blockchain is kept in a file which can be opened and copied by any node on the system. This gives it the decentralized feature.
2. Transparent: As the blockchain is an open file, anyone can open it and check the transactions. Hence, it can be traced over time.
3. Undisputable: A blockchain is a perpetual record of transactions. Once it is added, it cannot be changed, which brings trust in the record of the transaction.
4. Consensus Driven: Every single block on the blockchain is confirmed autonomously through agreed models that offer guidelines for authorizing a

block and regularly refer to a limited resource to display resilience that is sufficient to the work that was done.

2 Circular Economy (CE)

Circular economy is considered a substitute to a customary linear economy, which can be classified as take, make and waste. It attempts to eliminate the waste, improve resources at the end of the product's life cycle and connects it back into the line of production. Hence, it drastically eliminates the force on the environment [113].

The attitude towards the circular economy is crucial as it brings new chances for development. The added values of this approach are as follows: elimination of the waste, cause a higher productivity level, convey more competitive economy, help the economy to recognize the issues related to the limited resources in the future and aid in decreasing the environmental influences of the creations and intakes [113].

2.1 The Concept of Circular Economy

The focal point for the recent debate regarding the circular economy concept is a challenging question considering whether the creation of the waste in fact is disruptive towards our line of production. Substitute methods including circular economy, closed cycle, efficiency in the usage of the resources, reuse and recycling follow the thought of an accountable usage of material, products, environment and resources. Even though they have recently gained power, yet, having a world with zero waste could merely be accomplished with a complete concept. This indicates considering methods like reuse, recycling of both energy and materials at every single level of the product life cycle to guarantee the environmental product design from the beginning with a recycling goal at the end [118]. The ultimate awareness of the circular economy has increased several streams and alternatives including changes in the approach, scope and the concept, which comprise the circular economy of the blue economy concept, the Ellen MacArthur Foundation, cradle-to-cradle and zero waste that significantly vary with regard to the viewpoints on bio-based sequences [22, 73].

Moving towards circular economy is connected with high prospects related to the economic and ecological benefits. Transferring to a further circular economic model guarantees a greater future for the economy through aiding in separation of the economic growth from the usage of the resources and its influences. It provides the outlook of the Sustainable Development that will be continuous. Researches highlight these benefits via four stages: resource deployment, the economy, the environment and the community benefits, namely the formation of new jobs [33].

Resource deployment benefits: enhancing resource safety and eliminating the reliance on imports

The circular economy has the ability to enhance efficiency of initially raw material utilized at the universe level. If the materials are well kept in excellent products or the waste is moved back to the industry as an excellent subordinate raw material, the circular economy can eliminate the industry's request for key raw materials. The lesser the request for key raw materials will aid in decreasing the reliance on imports. This will make value chains in several sectors of the industry to be less impactful by the changes in price in the global commodity markets and so as to the uncertainty of the source produced by the shortage or other factors such as geopolitical ones. The recent evaluations recommend that about 6–12% of the total material intake can be eliminated via recycling, avoiding the waste and also via eco-design strategies. The highest probable considering the current technologies is approximately at up to 17% [34].

Environmental benefits: less environmental influences

The entire separation of economic development and the quality of life from intake of resources and energy are the main objective of the European Union's resource efficiency policy. Strategies related to a circular economy add values to that objective through many different techniques, namely through bringing the elimination of waste as a priority and reuse under the waste category (European [34].

Economic benefits: chances for economic development and modernism

Moving away from a linear line of production and intake as per the take, make use and dispose approach can also provide noticeable chances to enhance competitiveness in many sectors in the European industry. The circular economy can save cost for many industries. As per the assessments done by Ellen Mac-Arthur Foundation, enhancing the movement in the industrial of compound durable with a standard life cycle can eliminate material cost of up to USD 630 billion, only in the EU. Furthermore, the circular economy can also provide an opportunity for creative methods, technology and business models that can generate an economic added value from scare resources, and this can aid the European industry through becoming stronger against exterior shocks and so as enhancing its international competitiveness [116].

Community benefits: towards sustainable consumer behaviour and new job creation

From community perspective, Europe can benefit from moving towards a circular economy. Community modernisms along with the cost elimination, recycling, reuse and other progresses bring in chances to build more sustainable forms of consumer behaviour and hence participate in the human health and consumer security. Moreover, circular economy can create new job opportunities in Europe. As per the European Commission's impact assessment for waste objectives, enhanced monitoring and distribution of best practices alone can generate more than 180,000 new jobs in 2030 [33].

Nevertheless, it is obvious that as of now, particular questions still have not received enough attention in the circular economy discussion. Regardless, energy is needed to recycle the waste. Even though, this is usually not highly required for mining and handling key raw materials, it continues to be difficult to circulate infinite amounts of materials without encountering the climate objectives [114]. Basically, the movement to the circular economy will not hinder the need to drastically limit the usage of natural resources in the benefits of Sustainable Development.

Another related feature is the readiness of raw materials. Till recently, the resource debate has been controlled by raw materials that are very crucial for specific procedures or products. There is no sufficient alternative for these materials, and simultaneously, supplies are scarce since demand is more than the supply, which could result in monopoly across small number of countries [32].

2.2 *Circular Economy Challenges*

According to ECO-INNOVATION at the heart of European policies [29], on average, some materials are aligned more easily with the circular economy as compared to others. Almost all foods, beverages, cans and other forms of steel packaging are reused. However, some other materials are more difficult to be recycled.

Fibre-reinforced plastic (FRP) is a combination of glass and plastic, carbon and other fibres used in many applications such as doors, wind turbine blades and vehicle machineries. From an environmental perspective, FRB provides major benefits since it is resilient, light and last for a long period of time. For instance, FRP machineries decrease the vehicle's weight, their consumption of fuel and therefore, their greenhouse gas releases [29].

FRP's environmental Achilles' heel, yet, is that it is difficult to reuse or recycle. Around 60% of the remaining FRP is landfilled. The rest of the FRP is destroyed. As it is recycled, it is usually used for lower applications, namely a filler in the new compound resources. Hence, it is difficult to generate income, and there is low economic enticement to recycle FRP [29].

Enhanced methods of reusing and recycling the material are profoundly needed since the amount of FRP created is increasing, accordingly the quantity of waste FRP will certainly increase as well. A huge project by European Union research has been functioning since 2017 to display that circular economy approaches can be implemented on massive products created from FRP. The ECOBULK project will determine circular idea regarding FRP in three scenarios: cars, equipment and buildings. The demo is predicted to occur in 2020 [29].

ECOBULK will continue running till mid of 2021. The study focuses on FRP through the entire development of bulky products, which starts from design, recycling and then reuse. It also focuses on observing the circular business models, namely sharing and leasing, and exactly how the logistic chains can be developed more sustainable for huge products [29].

One of the examples of the projects is new modular parts for car centre consoles, which can be easily disassembled at the end of car's life. As for the equipment, ECOBULK is reviewing new methods of gathering pieces that are made from FRP through eliminating the usage of toxic cleaners, which will make it easier for the equipment to recycle [29].

One of the ECOBULK work element is viewing how to handle the waste problem embodied by wind turbine blades. These wind turbines have a life cycle of about 30–40 years, and finding practices for neutralized FRP blades is going to be a major material test in the upcoming [29].

2.3 *Ecosystem Economy and the Cities' Role in Driving CE*

One of the core assumptions, which is supported by the result of the FES discussions, reveals that the debate on the circular economy in Germany is still focusing heavily on the waste management matter. The concentration remains on measures that only happens at the end of the lifespan of the product, such as improved separation of the material that are recyclable from the residual waste. One core challenge will be about sharing that the circular economy is broader than waste separation and improved management of the waste [118].

Other factors including the innovation in the technical side play an important role in the circular economy. This is crucial in the stage of designing the product which requires to be long-lasting, rectifiable, and most importantly, 100% recyclable. In fact, the technical side of the circular economy is an easy part in the challenge of shifting entirely to a circular economic system. All related parties, namely resource producers, consumers, suppliers and product designers, will have to work side by side on improved solutions, instead of focusing only on their elements of the chain, which could be enhanced recycling rates, process enhancement and improved resource extraction [118].

It is clear that the circular economy demands for a robust regulatory structure. The debate regarding the probable economic savings and market prospect portends to reduce the light on the fact that so many players in the market are benefiting from the current economic linear system. Moving towards circular economy will definitely not happen automatically, and the new business models will only be able to play a vital role in driving towards circular economy if they are granted with sufficient regulatory structure [118].

In order to build a structure that will aid in circular economy, new policy tools are required which is out of the current waste regulation.

The main test will be to incorporate these tools in a new policy mix:

- Where single features are balancing and preferably supporting. For instance, in Germany, related policy is yet seeming to be unpredictable, and most of the current measures are yet considered for a traditional linear system.

- Which groups sensible structure tasks that are spread across an extensive array of political parties and ministries. This covers the query of the accountability of the domestic experts and private-sector waste workers, which needs to be taken into consideration from the view of a long-run circular economy and less with regard to the short-run market share.

Such policy mix is essential to build a long-term steady and reliable structure where businesses would invest in advanced circular-capable production method and users will be able to benefit from the advantages of this sustainable economic model [118].

To sum up, the expert conversations have shown the latest action of this concern and vividly illustrated that despite the predictable part of the ecologic and economic capabilities, the application mainly reveals major challenges [5].

2.4 *Circular City*

In the past, our method of production and consumption has added to the world-related concerns, and this includes inefficient resource allocation, inequality at the social level, weather conditions and the pollution in the ocean. It is predicted that these concerns will increase and extend to a crisis level [105].

The current economic system is allowing us to consume more sources than the world can replace [105], which is impacting the financial yields and the social and ecological side of the world. At this moment, it is crucial to reconsider the approach we use for operating and transiting to a better sustainable future [40] through 'Sustainable Development (SD)' [90]. This method broadens the horizon to an original and essential method of reaching equilibrium across economic well-being and social equity and simultaneously living within the capacity of the world.

Within the available methods, there is a circular economy (CE) model, which is a new economic and development model that has received a great emphasis in the last years [42, 58]. According to Ellen MacArthur foundation (EMF), CE is an economy-built base on the values of planning out pollution and excess, keeping the products and materials in practice and restoring natural methods (Ellen MacArthur Foundation 2019). Yet, this model still requires validation and evaluation [48, 107] in order to determine its capacity to be the most sufficient example for Sustainable Development.

Cities are moving the global economy forward. At the moment, around 55% of the world population are living in the cities, and it is predicted that this percentage will increase to 80% in 2050. Residents of these cities incorporate to about 80% of the world GDP. In addition, in order for the cities to sustain their ecosystems, they require about 40 billion tons of sources, and it is predicted to reach about 90 billion in 2050 [110]. Cities are the hub of consumers and businesses. Hence, cities' role in moving the circular economy forward and achieving sustainability in the upcoming has been counted as significantly valuable and needed [12, 69].

Cities provide the ideal scale and environment for circular economy when it comes to production and consummation of the sources, which aid cities to transmit to CE to consume the self-declared name 'circular hot spots' [91]. The European Union recognized all of these and endorses circular economy to face these challenges through devoting funds to implement it (Horizon 2020, LIFE) and also a part of the economic recovery strategy of the European Union long-run budget 2021–2050.

In 2019, the Asian Development Bank predicted that by 2025, among 37 of the globe's megacities, 21 of them with a population of 10 million or above, will most likely be in the Asia and Pacific region. 80% of the region's GDP comes from Asian cities, where they use about 60–80% of all energy and sources and generate 75% of the carbon releases of the region. Moreover, seven Asian countries, namely China, Japan, Indonesia, Thailand, Malaysia, India and South Korea, contribute to about 45% of the global GDP in 2050 [4]. It is also predicted that during the period 2015–2030, cities will be responsible for 91% of global consumption growth [74].

There is a need to understand the basis of a circular city, its structure that takes into account the local framework of where it will be applied.

Circular city is the city that applies CE principles to near resource circle in association with its interested parties to achieve a future-proof city ([91], p. 187).

The list of cities that demonstrate direction and movement towards circular economy are Amsterdam [94], Glasgow [18], Rotterdam and Charlotte [45, 46]. Furthermore, other cities, namely Porto [36] and Bilbao [17] are on their path to become circular cities. They use current data to plan the shift towards a 'fully circular city' which is in line with their domestic circumstances and therefore introducing new measurement, strategies and guidelines along with tools to track the values that can be raised with the circular shift. So far, variety of models have been offered for cities applying circular economy [70, 91, 111].

So far, there is no structure that has been implemented, examined and proved to be operational. However, there are three methods that are worth highlighting here taking into account their direction towards their applications, an idea for a circular city the ReSOLVE framework (2) and the circular city analysis framework (CCAF) (3).

1. A study conducted by circle economy and Holland Circular Hotspot [95] recommends a transition from linear to circular, in particular percentages for the following classifications that collectively function in a city and therefore building a circular city as shown in table below. The method focuses on applying technology, design, social innovation and coalition building in a collective method and examines the methods main businesses and sectors can be shifted to circular along with the general advantages a city can obtain through moving to a circular structure.
2. The ReSOLVE framework was presented by the [30] and recognizes these six pillars that businesses should apply for a shift towards circular economy. Such method was not initially planned for CE application in cities yet some of its principles add towards a circular city as long as they are implemented thoroughly. One of the pillars 'Regenerate' aims to reinstate the natural capital and

escalate the ecosystems' flexibility through recurring the precious biological nutrients carefully to the environment either through anaerobic absorption or composting. This is facilitated through the support to support viewpoint which indicates that technical and biological beneficial need to be dispersed during the product's lifespan from the early design stage onwards. In the constructed environment, this pillar indorses the use of renewable energy to influence construction. On the other side, the 'Share' pillar follows the extreme use of modules, products or resources during sharing systems or give and take platforms. As for the 'Optimize' pillar, it is more related to improving the efficiency and the functioning of the product, automation and distance sensing and so as eliminating the excesses in the production. Regarding the 'Loop' pillar, it has four core goals, which are as follows: to reuse material, to remove biochemical from biological excesses, to abstract excesses anaerobically and to obtain the production of modules and products. One of the other pillars 'Virtualise' concentrates on two kinds of dematerialization, which are direct and indirect dematerialization. The preceding pillar is 'Exchange' pillar which has three main categories including new technologies; innovation by choosing new goods or services; or the replacement of old materials with new materials [16].

3. The latest published structure is the circular city analysis framework (CCAF), which focuses on applying and tracking CE revealed through the literature reviews. This structure is already used for tracking objectives in the city of Porto, Portugal [16] (Tables 1 and 2).

2.5 Sustainable Development

Circular economy is considered a core strategy in accomplishing the Sustainable Development and the United Nations Sustainable Development Goals (SDGs). To grasp the perspective of CE, organizations should value and communicate their movement away from the non-sustainable model of 'take-make-dispose' towards circularity. Both literature and experts display a strong interest in the assessment of CE as a driver for this change. However, most of the CE valuations are taking place in private-sector companies, and hence, the execution at the public sector is still low. A study was conducted where around 21 CE and valuation specialist from the public sector of Portugal were interviewed [28].

The result of the study shows that cultural constraints, mainly the absence of political and public forces along with the struggle against change, are the core challenge for executing CE valuation. These cultural challenges influence the basic ones, namely the absence of commitment towards leadership, the charitable nature along with the absence of governance for CE valuation. On the other side, financial and technical challenges are not priorities and are mainly the result of the cultural and basic challenges [28].

The result of the study recommends that CE valuation is regularly a debate across academics and experts. In order to speed up its execution, the debate must

Table 1 Vision for a circular city [95]

Housing and infrastructure	Buildings globally account for 45% of global resource consumption. A circular city is literally built with renewable, non-virgin and low-carbon footprint materials
Mobility	Transportation sharing, and renewable and clean fuel will drive circular mobility for cities, as cities account for 40% of all transport-related emissions
Food	Cities are expected to consume 80% of all food by 2050 and the worldwide food system is responsible for 20–30% of GHG emissions. A circular food system will focus on locally produced food, minimising food waste by prevention and repurposing of generated waste
Energy	Already, 75% of worldwide energy consumption takes place in cities. Renewable energy will fuel the circular city by hyper-local, decentralised grids. Energy loss is prevented and energy generated in access, captured
Water	A circular city minimises extraction and pollution of local water-ways and uses closed loop systems for its water flows; resources are recovered from wastewater
Consumer goods	Circular design will offer a completely different approach to production and consumption, monetised by circular business models
Plastic	A circular city bans traditional single use plastics. New materials or traditional materials are adopted and landfill, incineration or any contribution to the plastic soup is prevented by policy and lifestyle
Industrial parks	Circular Industrial Parks are driven by eco systemic functions, symbiosis and the use of waste as a feedstock

include other related parties outside the expert loop to increase the level of awareness regarding how far important CE is and to ease the practice of it for a larger audience [28].

CE valuation is getting attention as an approach to move the shift from the non-sustainable model of take, make and dispose to the recurring use of the sources and development of used products and probably participating in the Sustainable Development [56].

It is considered among the vital strategy to a more sustainable model and in accomplishing the United Nations Sustainable Development Goals (SDGs) (Schroeder et al. 2019). While moving towards CE, assessing the circularity of goods, services and strategies or their role in the movement towards CE is important for planning and highlighting circular results as per the data [23]. Hence, a group of CE valuation methods has been established in past years.

Roos [23, 37, 67] established, inventoried and evaluated many different CE valuation methods. As a result of a lack of recognized description of the CE notion, the current CE valuation methods assess circularity in diverse ways [61, 96]. The theory reveals that majority of the current CE valuation methods emphasize on assessing the efficiency of the resource along with the flows and stocks of the material [77, 88].

Table 2 Singapore and Porto comparative table using the CCAF

Field	Indicator	Singapore	Porto
Local resources	1. Wind potential (m/s)	3.98	6.78
	2. Solar potential (W/m^2)	1652	1750
	3. Green roofs (%)	0 34	0 50
	4. Import/exports (%)	0.9	1.5
Renewable energy	5. Renewable energy penetration (%)	8	63
	6. Access to electricity (%)	100	100
CE innovation	7. CE Innovation budget	0	0.009
Food	8. Food waste treated (%)	17	21
Transport	9. Public transport usage (%)	60	19.6
	10. Electrical energy use in transportation (%)	5.5	0.6
Recycling	11. Recycling rate	60	100
Water management	12. Safe water accessibility (%)	100	100
Waste management	13. Landfilled waste (%)	2	1
Education	14. Basic education quitting (%)	6	11
Digitalisation	15. Accessibility to smartphones (%)	90	71.6
Demographic	16. Balance between women & Men (%)	42	55
	17. Heaviest age group (years)	45–54	60–69
Policies	18. Active population (%)	67.7	59.2
	19. Women & men balance in Politics %	23	38

Source Ferreira and Fuso-Nerini [36]

Other methods also measure the influence of CE implementations and strategies on Sustainable Development [59, 72]. It is vital to measure the influence of CE activity on Sustainable Development as CE implementations do not inevitably result in an enhancement in the Sustainable Development [11].

The CE valuation methods focus on three levels: micro, meso and macro. Micro-level focuses on organizations, goods and material, while meso-level targets industrial networks. As for the macro-level, it focuses on rules and policies [58, 62]. Both public and private sectors play key roles in the movement towards CE [6]. The application and studies of CE valuation are mainly focusing on companies in the private sector [38]. Nevertheless, public sector has different missions, and it is planned as per the bureaucratic values [65, 117].

The bureaucratic values impact the practices and sustainment of the valuation methods [115]. As on date, the participation of the public sector to the CE is mainly evaluated and measured at the macro-level. Back in 2008, one of the first countries that announced a particular structure of indicators to monitor the progress was China. The law was called 'Circular Economy Promotion Law' [41, 89]. This law was the kickoff of series of attempts in building suitable indicators to value CE policies [19].

At the micro-level, public sector plays an important role in shifting towards Sustainable Development [102]. The organizations under public sector are the key economic players as the public procurements in the EU are around 14% of the GDP and considered as a good example to other public-sector bodies, citizens, NGOs and so as the private sector [27, 35]. Public-sector bodies have already started applying the CE and sustainability implementations and strategies in their processes, namely procurement activities [13, 63].

It has been over decades since public-sectors bodies have begun adding performance evaluation method in their managerial implementations and processes [50, 51]. This model moved from the customary bureaucratic bodies to take into consideration more of the managerialism, which is called new public management (NPM). In this model, the valuation is considered as the main topic for the public managers [49]. The notion of 'more with less' has turn into a motto where the public managers aim to sustain or develop the quality of their service distribution [2].

However, the link between the CE activities and valuation has not been made yet. Despite the fact that most of the public-sector bodies have stated their likelihood for evaluating their CE performance, its application is yet at a beginning level [42]. In order to grasp more of the CE perspective, it is important to apply the CE valuation in the processes and strategies [60].

3 Application Scenarios of Blockchain Technology for a CE

Blockchain technology is a new practice of sharing and updating information through connecting databases in a decentralized, open-access, peer-to-peer network. It has been developed to guarantee the data is kept and updated in a safe, permanent and temper-proof method. Even though blockchain is still at its emerging level, the research in this area has been rapidly growing, which makes it vital to take into account the moral and the sustainability effects of blockchain progress and application. CE also emphasizes on improving sustainability and social accountability beside economic development [3].

Blockchain technology can participate in the CE through aiding in the reduction of the transaction cost, improve performance and communication within the supply chain, guarantee human right safety, increase healthcare patient privacy and prosperity and decrease the carbon footprint [3].

The CE assures helping the development of a constructive human future through the usage of technology and smart design to make the most of the use of the sources and decrease the waste [71].

One way to make the CE effective is through giving a purpose to the new approaches, which could be sustainable resource creation and intake, reproducing the products and recycling. Another approach could be ensuring that the recycled goods that companies and people are buying are not made from raw materials. In the absence of transparency and confidence, it is probable to return back to the current linear economy. Blockchain has the capability to develop that confidence [71].

Blockchain is not as difficult as it may sound. It is a circulated record that eliminate the need for a trusted third party to check the transactions. This technology will keep the information on transactions in present time in 'blocks' that are cryptographically connected to generate an absolute record of activity. The design of the blockchain assists in two main usages for the CE, namely verifying the origins of the product and motivating constructive behavioural adjustment [71].

The CE indorses sustainable management of the sources, reducing the waste and recycling of the resources. To connect these applications and actions, there is a need for transparency to handle confidence in the product we purchase and who we are purchasing from. Higher transparency in terms of the product origin will aid in constructive intake and push companies to adjust the way they obtain sources. Blockchain is considered a foundational technology for the building of clear digital supply chain, which provides an absolute record of transaction that audit the root of the products [71].

Promoting behavioural change without a robust and direct motivation can be challenging. This is more or less the same as when attempting cultural changes required to shift businesses and society towards a circular economy. One of the approaches to promote CE would be rewarding circular buying, usage or disposal. Blockchain has the ability to tokenise natural resources, providing them with a unique digital identity where people can use while trading. This method makes the value of the resources more obvious, enabling a new mechanism of setting the price and trading of the natural resources. It also motivates individuals to embrace circular behaviours [71].

Blockchain stocks the data, and this could be data from IoT devices or RFID tags. This connection with the physical world generates challenges, specifically with regard to the quality of data [71].

IBM and Maersk have built a global supply chain solution constructed on blockchain that already attempt to sell 10 million supply chain events per week [71].

Blockchain can possibly turn the CE into an advanced ecosystem through providing product assurance and dynamic boosts. The current stage is more likely a testing stage, however as soon as large-scale pilots demonstrate the significance of the blockchain, a huge acceptance will follow. Societies will have a chance to discover and examine the technology via proof of ideas and pilots today. Taking into consideration the capability of blockchain along with the fast pace of the advancement in the technology, this is the right time to begin testing with it [71].

3.1 Focus on Singapore

Singapore has always been conscious towards having a balance economic development and environmental sustainability. For the economy to grow, water, energy and other sources are required, on the other side it creates sewerage and solid waste which is needed to be disposed of. To handle these challenges, it is important for

Singapore to adopt a circular economy method, which carries a holistic shift from use and through mentality to reuse and recycle for as long as possible [92].

Singapore is in the process of presenting policies that support sustainable production and intake, which covers the extended producer responsibility (EPR), that begin with electronic waste (e-waste). EPR will not be limited to e-waste, and it will also cover the packaging waste as well. Once the producers are accountable for the end-of-life of their goods, they will have further motivation to create goods that can be easily recycled, or build an innovative circular business prototypes to limit their waste loop [92].

Lately, a research funding under the title 'Closing the Waste Loop Initiative' was established for designing sustainable plastic materials. The objective is let plastics to be more recyclable, easier to reuse and to remove value from the waste plastics [92].

The research findings show that there is still a shortage of a vigorous circular cities case studies, and there is an absence of an agreement towards suggesting a model to be accepted and applied [36]. A thorough review over the Singapore's method to circular economy is currently not available.

Looking forward, Singapore is a model in terms of how the cities that are to be circular are emerging. Singapore's practices on the circular economy are different from other developing countries, and these practices bring insights into the design and application of circular economy policies for cities with similar characteristics [26]. In spite of the individuality of Singapore as a small island state, it has apparently different performances. As per the World Wildlife Foundation (WWF) report, the country has the seventh largest ecological footprint in the globe [75]. This is a very high influence taking into account the country's economy rank in terms of GDP, which was 37 as per the International Monetary Fund [55].

The country is also known for being the Southeast Asia's most modern city with regard to development of buildings and commercial constructions [66]. Furthermore, some of the sustainable innovation started in Singapore, mainly in the field of transportation, have already been implemented in other cities, namely Shanghai, London and Stockhom [26]. Several enterprises come under the big vision of the 'City in a Garden', which goes back to 1963 [100], or the upcoming vision, which is about 'A City in Nature' as lately issued via Singapore's Centre for Liveable Cities, a government organization that is a part of the Ministry of National Development [99].

Nearly, all of the Singapore city–state is less than 30 years old and is handled in a way where the uncertainty is omitted. According to the well-known architect Koolhaas [103], if one come across a chaos in Singapore, then it is probably deliberate, 'even nature is entirely remade'. The whole story of the country explains how it handled difficulties to be shifted from a colonial backwater into a thriving metropolis, for instance, in international studies that focus on innovation rate, Singapore is always on top of the list of the globe's most innovative economies [43]. Even though everything is recreated and planned, accepting the process related to circular city is still a challenge.

The country has adopted the year 2019 as the year towards zero waste [113], as Singapore is operating towards becoming a zero waste country via decreasing its intake of materials and growing reprocessing and recycling rates.

In 2019, the country's zero waste masterplan was issued, the plan focuses on three waste streams, namely e-waste, packaging waste and food along with new technologies and innovation that the government is discovering to fully eliminate or close the waste loops. At the moment, comprehending these streams depends upon the recycling rates [100] along with the focus studies for recycle.

During the same period in 2019, the Resource Sustainability Bill was released. As soon as it gets approved, it will 'impose obligations relating to the collection and treatment of electrical and electronic waste and food waste, to require reporting of packaging imported into or used in Singapore, to regulate persons operating producer responsibility schemes, and to promote resource sustainability' ([101], p. 1).

However, the latest issue of the country's Economic Budget for the year 2020 [98] does not reveal any symptoms of circular economy, and there is a marginal focus on sustainability economic policies during COVID-19 pandemic. This is different as compared with Europe and China where the circular economy is expanding which shows in the policy and their annual budgets that target maintaining the economic growth and simultaneously enhancing social equity and the quality of the environment.

Yet, in past years, the country has taken several steps towards a circular economy which is worth mentioning. Some of these efforts are as follows: prominent stakeholders, namely JTC, the industrial land authority that is studying how the industrial zone 'Jurong Islan' may become more circular, began exploring the CE model [76]. Also, one of the latest partnerships with Neste, which is the world's leading supplier of renewable diesel, will permit Singapore to lead the creation of this fuel and also jet fuel by 2022 [112], via altering raw materials such as waste animal fat, plant oils and possibly dissolved waste plastic to renewable energy. Singapore is responsive, and once it emphasizes on an idea, it will expand into a prominent force (Fig. 2).

Majority of the country's zero waste agenda have been built lately via latest governmental attempts (e.g. Singapore's long-term low emissions development strategy) and other regulations that are scheduled for the upcoming 5 years to accomplish the country's main objective of 30% decrease in the waste by 2030 along with the 70% growth in the recycling of household. Moreover, financial incentives have also been presented to motivate private sector towards a circular economy [82].

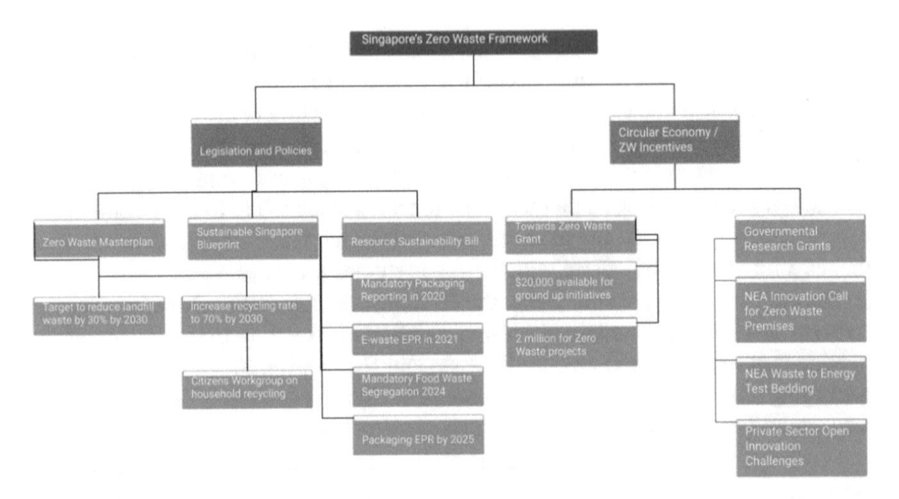

Fig. 2 Singapore's zero waste framework [16]

4 Challenges of Applying the Blockchain Technology to the CE

Lately, blockchain has received a massive attention from scholars, engineers and institutions who will benefit from blockchain technology. Their applications have influenced fields like e-finance, e-health care, e-government, smart contracts, logistics and so on [24].

A new form of virtual currency was developed by Satoshi Nakamoto and is called cryptocurrencies. These currencies are more protected and confirmed method of transaction and rely on cryptographic protocols in a peer-to-peer structure, which generates a distributed ledger. As blockchains started to become more and more popular, a huge number of cryptocurrencies have been presented, which is highly changing the visions of blockchains in the finance field. In these blockchain structures, all the dedicated transactions are kept in a list of blocks with many transactional assets [24].

Since both consent and unconsent blockchain platforms are facing difficulty to meet the demanding real-world implementation requirements, namely low latencies, high performance, good scalability and immediate transaction finality, the limitation of the current model is being identified. There are several new models which have been designed and still are developing to handle the limitation of their prototypes [24].

Even though the future of the blockchain in nurturing the circular economy is quite promising, yet the pathway has room for further improvements. There are several challenges that one has to take into consideration during the implementation of its application. The contradiction of synchronized opposition and collaboration [10] inside a blockchain in the circular economy can increase the matters related to conflict, confident and co-dependency [80]. The automation side of the smart contracts has also become a source of a concern amongst auditors, accounts and lawyer along with the future of their roles [53, 97]. The supporting philosophy here is to shift from the transactional to the relational and plan for business models that are reinforced by experts, through emphasizing more on social welfare and societal than only profit-oriented markets [81]. Blockchain intends to prompt the confidence of the stakeholders via its distributed and immutable algorithmic structure [31]. Hence, blockchain performs as the reliable third-party authenticator of transactions. Since all the archived are kept and restructured with unique nodes of the network, the responsibility of the confidence moves from a third-party broker, accordingly, offering safety and avoiding interfere [97]. Perhaps, as confidence in the structure is improved via usage, blockchain technology will build endurance and add respected aids to the circular economy via sustainability and social responsibility [3].

5 Suggestions for Way Forward

There exist a complex link and massive possibility of blockchain technology to positively contribute to the circular economy via sustainability and social responsibility plan. Blockchain is supporting applications in many sectors via its ability to enhance tracking. Blockchain implementation is analysed in several sectors and how it is aligned with social responsibility and sustainability matters. The analysis reveals that blockchain plays a role in the circular economy via its decentralization, ability to distribute with high security features. Moreover, the same analysis reveals that blockchain participates in the circular economy through aiding to decrease the transaction cost, improves performance and communication along the chain of supply and makes sure that the human rights are protected. Furthermore, it improves the confidentiality and welfare of the patients in the health care and decreases the carbon footprint. There is a possible upfront cost intricate in the implementation of blockchain technology, even though the benefits are more likely to outweigh the challenges. According to a study, a narrative review approach was adopted to collect two relatively new fields of study. Yet, taking into consideration the exponential rate at which the studies related to blockchain are prepared and published, more reviews would benefit from a more quantitative perspective. There are several articles examining how blockchain technology can play a role in the circular economy. Furthermore, there is a shortage of studies emphasizing on social responsibility within the circular economy model. In addition, regulation in diverse countries is most probably going to impact the growth and implementation of the blockchain. More studies on the policies and regulatory implication for blockchain and circular economy development along with studies related to developing countries with challenges in their infrastructures are recommended [3].

It is expected that the future cities function differently from the past cities, and circular economy would definitely aid in this shift. Singapore has many chances for growing towards circular economy, taking to account its initiatives towards Sustainable Development. In fact, the country has the ability to become leading circular

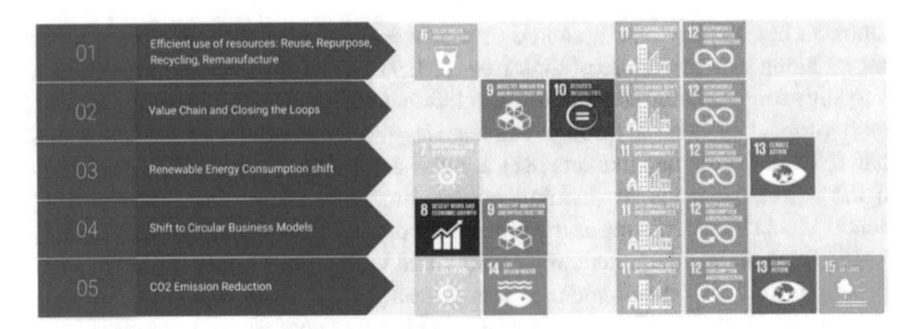

Fig. 3 Circular economy and sustainable development goals [16]

city. Singapore's government wide vision and plan on the circular economy allow businesses to apply the needed transitions, and this is predicted to result in a more demanding application of circular values [16].

The country can begin by building a physical place where an innovating community can join and exchange their recent developments. Such 'Circular Hub' would be continually creating, giving a chance to start-ups to catch the testing ground for their circular ideas and goods [16].

There is a vital demand for more study on what the circular city method may carry for the country's system of food, energy, agility, water, plastic and industrial grounds (OECD 2019). Furthermore, to participate in addressing circular economy as the most appropriate mode for Sustainable Development, more research is required on how the Sustainable Development Goals are relevant to a circular economy for Singapore [16] (Fig. 3).

Even though each city has diverse needs, priorities, obstacles and challenges, setting one single pathway towards circular economy seems to be impossible.

6 Conclusion

This chapter aimed to provide a better understanding of the definition and characteristics of blockchain. Then, several application scenarios of blockchain technology for a CE with focus on Singapore while transitioning to a CE were explained. The country has numerous opportunities for growing towards circular economy. In fact, Singapore has the capability to turn into a leading circular city. The country's wide vision and plan on the circular economy allow businesses to apply the desirable transitions, which is predicted to result in a more demanding application of circular values. Singapore can start with building a physical place where an innovating community can join and exchange their recent developments. Such 'Circular Hub' would be continually creating, giving a chance to start-ups to catch the testing ground for their circular ideas and goods.

Furthermore, some of the challenges that still exist when applying the current blockchain technology to the CE have been discussed and analysed. It is worth noting that despite the fact that the future of the blockchain in nurturing the circular economy is quite promising, yet the approach has room for further improvements. As discussed, some of the main challenges that need to be taken into consideration during the implementation are the contradiction of synchronized opposition and collaboration inside a blockchain in the circular economy, which can increase the matters related to conflict, confident and co-dependency. Moreover, it is predicted that the future cities function differently from the past cities and circular economy would definitely aid in this shift.

There is a strong demand for further studies on what the circular city method may carry for the country's system of food, energy, agility, water, plastic and industrial grounds. Moreover, to contribute in addressing circular economy as the most appropriate mode for Sustainable Development, more studies are required on how the Sustainable Development Goals are relevant to a circular economy for Singapore.

Lastly, while each city has different requirements, priorities and challenges, allocating one single approach towards circular economy seems to be unfeasible.

References

1. Alt R, Beck R, Smits MT (2018) FinTech and the transformation of the financial industry. Electron Mark 28(3):235–243
2. Arnaboldi M, Lapsley I, Steccolini I (2015) Performance management in the public sector: the ultimate challenge. Financ Account Manag 2015(31):1–22
3. Arvind U, Sumona M, Vikas K, Yigit K (2021). Blockchain technology and the circular economy: implications for sustainability and social responsibility. J Clean Prod 293. ISSN 0959-6526, https://doi.org/10.1016/j.jclepro.2021.126130. Accessed 1 May 2021
4. Asian Development Bank (2019) ADB's vision of livable cities. https://www.adb.org/themes/urban-development/adb-vision-livable-cities. Accessed 1 May 2021
5. Bahn-Walkowiak B, Wilts H (2015) Reforming the EU VAT system to support the transition to a low-carbon and resource efficient economy. In: Kreiser L (ed) Carbon pricing—Design, experiences and issues. Chel-tenham, Elgar, pp 111–126
6. Barriers to the Circular Economy (2018) Evidence from the European Union (EU). Ecol Econ 2018(150):264–272
7. Beck R, Avital M, Rossi M, Thatcher JB (2017) Blockchain technology in business and information systems research. Bus Inf Syst Eng 59(6):381–384
8. Beck R, Czepluch JS, Lollike N, Malone S (2016) Blockchain—The gateway to trust-free cryptographic transactions. In: Proceedings of the 24th European conference on Information Systems, Istanbul, Turkey, pp 1–14
9. Beck R, Müller-Bloch C, King JL (2018) Governance in the blockchain economy: a framework and research agenda. J Assoc Inf Syst 19(1):1020–1034
10. Bengtsson M, Raza-Ullah T, Vanyushyn V (2016) The coopetition paradox and tension: the moderating role of coopetition capability. Ind Market Manag 53:19e30. https://doi.org/10.1016/j.indmarman.2015.11.008
11. Blum N, Haupt M, Bening C (2020) Why "Circular" doesn't always mean "Sustainable." Resour Conserv Recycl 2020:162
12. Bonato D, Orsini R (2017) Sustainable cities and communities design handbook, 2nd edn., Clark W (ed). Butterworth-Heinemann, UK, pp 235–245

13. Brammer S, Walker H (2011) Sustainable procurement in the public sector: an international comparative study. Int J Oper Prod Manag 2011(31):452–476
14. Buterin V (2015) On public and private blockchains. Ethereum Fuondation Blog. https://blog.ethereum.org/2015/08/07/on-public-and-private-blockchains/. Accessed 1 May 2021
15. Buterin V (2016) What are smart contracts? A beginner's guide to smart contracts. Blockgeeks. https://blockgeeks.com/guides/smart-contracts/. Accessed 1 May 2021
16. Carrièrea S, Rodríguezb R, Peya P, Pomponib F, Ramakrishnac S (n.d.) Circular cities: the case of Singapore. https://www.napier.ac.uk/~/media/worktribe/output-2671279/circular-cities-the-case-of-singapore.pdf. Accessed 2 March 2021
17. Circle Economy (2018) Circular Bilbao and Bizkai. https://www.circle-economy.com/wp-content/uploads/2018/04/Circular-Bilbao-.pdf
18. Circle Economy (2016) Circular glasgow. https://www.circle-economy.com/wp-content/uploads/2016/06/circular-glasgow-report-web-low-res.pdf
19. Circular Economy Indicators (n.d.). https://ec.europa.eu/environment/ecoap/indicators/circular-economyindicators_en
20. Coinbase (2017) What is the bitcoin blockchain? https://help.coinbase.com/en/coinbase/getting-started/crypto-education/what-is-the-bitcoin-blockchain#:~:text=The%20blockchain%20is%20a%20distributed,one%20bitcoin%20transaction%20to%20another. Accessed 1 May 2021
21. Coindesk (2017) A (Short) guide to blockchain consensus protocols—CoinDesk. https://www.coindesk.com/short-guide-blockchain-consensus-protocols. Accessed 1 May 2021
22. Connett P (2013) The zero waste solution. Chelsea Green Publishing, White River Junction. https://www.chelseagreen.com/product/the-zero-waste-solution/. Accessed 1 May 2021
23. Corona B, Shen L, Reike D, Carreón JR, Worrell E (2019) Towards sustainable development through the circular economy—A review and critical assessment on current circularity metrics. Resour Conserv Recycl 2019:151
24. Deepak S, Shrid P, Mehul S, Shikha B (2020) Cryptocurrency mechanisms for blockchains: models, characteristics, challenges, and applications. In: Handbook of research on blockchain technology. Academic Press, pp 323–348. ISBN 9780128198162, https://doi.org/10.1016/B978-0-12-819816-2.00013-7. Accessed 1 May 2021
25. Derks J, Gordijn J, Siegmann A (2018) From chaining blocks to breaking even: a study on the profitability of bitcoin mining from 2012 to 2016. Electron Mark 28(3):321–338
26. Diao M (2018) Towards sustainable urban transport in Singapore: policy instruments and mobility trends. Transp Policy 1–11
27. Domingues AR, Lozano R, Ceulemans K, Ramos TB (2017) Sustainability reporting in public sector organisations: exploring the relation between the reporting process and organisational change management for sustainability. J Environ Manag 2017(192):292–301
28. Droege H, Raggi A, Ramos T (2021) Overcoming current challenges for circular economy assessment implementation in public sector organisations. Sustainability 13(3):1182. https://doi.org/10.3390/su13031182. Accessed 1 May 2021
29. ECO-INNOVATION at the heart of European policies (2021) Eco-innovation action plan-European commission, "Tackling the toughest circular economy challenges". https://ec.europa.eu/environment/ecoap/about-eco-innovation/research-developments/tackling-toughest-circular-economy-challenges_en. Accessed 30 Mar 2021
30. Ellen MacArthur Foundation (2015) Growth within: a circular vision for competitive Europe. https://www.ellenmacarthurfoundation.org/assets/downloads/publications/EllenMacArthurFoundation_Growth-Within_July15.pdf. Accessed 1 Mar 2021
31. Engelhardt MA (2017) Hitching healthcare to the chain: an introduction to blockchain technology in the healthcare sector. Technol Innov Manag Rev 7(10):22e34. https://doi.org/10.22215/timreview/1111
32. Erdmann F, Schäuble N, Lang S, Jung M, Honigmann A et al (2011) Interaction of calmodulin with Sec61α limits Ca2+ leakage from the endoplasmic reticulum. EMBO J 30:17
33. Eur-lex.europa (2021) Legal content. http://eur-lex.europa.eu/legal-content/EN/TXT/?uri=CELEX%3A52014DC0398. Accessed 20 Mar 2021

34. European Commission (2020) The European semester. http://ec.europa.eu/europe2020/pdf/ags2012_annex1_en.pdf. Accessed 20 Mar 2021
35. European Commission (2017) European semester thematic factsheet: public procurement. https://ec.europa.eu/info/sites/info/files/file_import/european-semester_thematic-factsheet_public-procurement_en_0.pdf. Accessed 21 Mar 2021
36. Ferreira A, Fuso-Nerini F (2019) A framework for implementing and tracking circular economy in cities: the case of porto. Sustainability 11(6):1–23
37. Figge F, Thorpe AS, Givry P, Canning L, Franklin-Johnson E (2018) Longevity and circularity as indicators of eco-efficient resource use in the circular economy. Ecol Econ 2018(150):297–306
38. Fortunati S, Martiniello L, Morea D (2020) Strategic role of the corporate social responsibility and circular economy in the cosmetic industry. Sustainability 2020(12):5120
39. Fridgen G, Lockl J, Radszuwill S, Rieger S (2018) A solution in search of a problem: a method for the development of blockchain use cases. In: Proceedings of the 24th Americas conference on information systems, New Orleans, pp 1–10
40. Geissdoerfer M, Savaget P, Bocken N, Hultnik EJ (2017) The circular economy: a new sustainability paradigm? J Clean Prod 143(1):757–768
41. Geng Y, Fu J, Sarkis J, Xue B (2012) Towards a national circular economy indicator system in China: an evaluation and critical analysis. J Clean Prod 2012(23):216–224
42. Ghisellini P, Cialani C, Ulgiati S (2016) A review on the circular economy: the expected transition to a balanced interplay of environmental and economic systems. J Clean Prod 114:11–32
43. Gin B (2017) Singapore's long game in innovation. The Straits Times. https://www.straitstimes.com/opinion/singapores-long-game-in-innovation. Accessed 1 May 2021
44. Gladek E, Exter P, Roemers G, Schlueter L, Winter J, Galle N, Dufourmont J (2018) Circular rotterdam. Metabolic. https://www.metabolic.nl/projects/circular-rotterdam/. Accessed 1 May 2021
45. Gladek E, Kennedy E, Thorin T (2018) Circular charlotte. Metabolic. https://www.metabolic.nl/projects/circular-charlotte/. Accessed 1 May 2021
46. Glaser F (2017) Pervasive decentralisation of digital infrastructures: a framework for blockchain enabled system and use case analysis. In: Proceedings of the 50th Hawaii international conference on system sciences, pp 1543–1552
47. Hobson K, Lynch N (2016) Diversifying and de-growing the circular economy: radical social transformation in a resource-scarce world. Futures 82:15–25
48. Hood C (1991) A public management for all seasons? Public Adm 1991(69):3–19
49. Hood C (1995) Contemporary public management: a new global paradigm? Public Policy Adm 1995(10):104–117
50. Hood C (2004) Gaming in targetworld, targets in british public services. Public Adm Rev 2004(66):515–521
51. Hyvärinen H, Risius M, Friis G (2017) A blockchain-based approach towards overcoming financial fraud in public sector services. Bus Inf Syst Eng 59(6):441–456
52. Iansiti M, Lakhani KR (2017) The truth about blockchain. Harv Bus Rev 95(1):118–127
53. International Monetary (2018) World economic outlook. https://www.imf.org/en/Publications/WEO/Issues/2018/09/24/world-economic-outlook-october-2018. Accessed 1 May 2021
54. Kalmykova Y, Sadagopan M, Rosado L (2018) Circular economy—From review of theories and practices to development of implementation tools. Resour Conserv Recycl 2018(135):190–201
55. Kiayias A, Russell A, David B et al (2017) PPCoin: peer-to-peer crypto-currency with proof-of-stake. In: Proceedings of the 2016 ACM SIGSAC conference on computer and communications security—CCS'16 1919, pp 1–27. https://doi.org/10.1017/CBO9781107415324.004
56. Kirchherr J, Reike D, Hekkert M (2017) Conceptualizing the circular economy: an analysis of 114 definitions. Resour Conserv Recycl 127:221–232

57. Kjaer LL, Pigosso DC, McAloone TC, Birkved M (2018) Guidelines for evaluating the environmental performance of product/service-systems through life cycle assessment. J Clean Prod 2018(190):666–678
58. Klein N, Ramos TB, Deutz P (2020) Circular economy practices and strategies in public sector organizations: an integrative review. Sustainability 2020(12):4181
59. Korhonen J, Nuur C, Feldmann A, Birkie SE (2018) Circular economy as an essentially contested concept. J Clean Prod 2018(175):544–552
60. Kristensen HS, Mosgaard MA (2020) A review of micro level indicators for a circular economy & Moving away from the three dimensions of sustainability? J Clean Prod 2020:243
61. Kristensen HS, Mosgaard MA, Remmen A (2021) Circular public procurement practices in danish municipalities. J Clean Prod 2021:281
62. Lacity MC (2018) Addressing key challenges to making enterprise blockchain applications a reality. MIS Q Exec 17(3):201–222
63. Lazzini S, Anselmi L, Schiavo LL, Falanga AM (2014) The role of information systems to support performance management in public administration: the case of the Italian regulatory authority for the energy sector. Digit Transform Hum Behav 2014(6):47–64
64. Liew K (2018) Manifestos for sustainable development: sustainable modular steel-precast concrete building construction system for dwellings in singapor. In: Filho WL (ed) World sustainable series. Springer, pp 477–491
65. Lindgreen ER, Salomone R, Reyes T (2020) A critical review of academic approaches, methods and tools to assess circular economy at the micro level. Sustainability 2020(12):4973
66. Lindman J, Rossi M, Tuunainen V (2017) Opportunities and risks of blockchain technologies in payments—A research agenda. In: Proceedings of the 50th Hawaii international conference on system sciences, pp 1533–1542
67. Lorbach D, Wittmayer JM, Shiroyama H, Fujino J, Mizuguchi S (2016) Governance of Urban Sustainability Transitions, European and Asian experiences. Springer, Japan
68. Meulder MJ, De B (2018) Interpreting circularity. Circular city representations concealing transition drivers. Sustainability. 10(5):1–2
69. Mark L, Nic C, Patrick A (n.d.) Blockchain can drive the circular economy. https://www.paconsulting.com/insights/blockchain-can-drive-the-circular-economy/. Accessed 1 May 2021
70. Matschewsky J (2019) Unintended circularity? —Assessing a product-service system for its potential contribution to a circular economy. Sustainability 2019(11):2725
71. McDonough W, Braungart M (2002) Cradle to cradle: remaking the way we make things. North Point Press, New York
72. Mckinsey & Company (2016) Mckinsey global institute: urban world, the global consumers to watch. https://www.mckinsey.com/global-themes/urbanization/urban-world-the-global-consumers-to-watch/~/media/57c6ad7f7f1b44a6bd2e24f0777b4cd6.ashx. Accessed 1 May 2021
73. McLellan R, Iyengar L, Jeffries B, Oerlemans N (eds) (2014) Living planet report 2014, WWF. https://www.worldwildlife.org/pages/living-planet-report-2014. Accessed 1 May 2021
74. Witteveen+Bos (2019) JTC, Witteveen+Bos and Metabolic start study to explore industrial symbiosis on Jurong Island. https://www.witteveenbos.com/news/jtc-witteveen-bos-and-metabolic-start-study-to-%20explore-industrial-symbiosis-on-jurong-island/. Accessed 1 May 2021
75. Moraga G, Huysveld S, Mathieux F, Blengini GA, Alaerts L, Van Acker K, De Meester S, Dewulf J (2019) Circular economy indicators: what do they measure? Resour Conserv Recycl 2019(146):452–461
76. Mougayar W (2016) The business blockchain: promise, practice, and application of the next Internet technology. Wiley
77. Nakamoto S (2008) Bitcoin: a peer-to-peer electronic cash system. Bitcoin.org 9. https://doi.org/10.1007/s10838-008-9062-0
78. Narayan R, Tidstrom A (2019) Circular economy inspired imaginaries for sustainable innovations. In: Bocken N, Ritala P, Albareda L, Verburg R (eds) Innovation for sustainability. Springer International Publishing, pp 393e413. https://doi.org/10.1007/978-3-319-97385-2_21

79. Narayan R, Tidstrom A (2020) Tokenizing coopetition in a blockchain for a transition to circular economy. J Clean Prod 263:121437
80. National Climate Change Secretariat Strategy Group, Prime Minister's Office (2020) Charting Singapore's low-carbon and climate resilient future. https://www.nccs.gov.sg/media/publications/singapores-long-term-low-emissions-developmentstrategy. Accessed 20 May 2020
81. Notheisen B, Cholewa JB, Shanmugam AP (2017) Trading real-world assets on blockchain: an application of trust-free transaction systems in the market for lemons. Bus Inf Syst Eng 59(6):425–440
82. Notheisen B, Hawlitschek F, Weinhardt C (2017b) Breaking down the blockchain hype towards a blockchain market engineering approach. In: Proceedings of the 25th European conference on information systems, Guimaraes, Portugal, pp 1062–1080
83. Ostern N (2019) Blockchain in the IS research discipline: a discussion of terminology and concepts. In: Institute of applied informatics at University of Leipzig 2019. https://doi.org/10.1007/s12525-019-00387-2. Accessed 1 May 2021
84. Oxford Dictionaries (2018) Blockchain | definition of blockchain in English by Oxford Dictionaries. https://www.coindesk.com/oxford-dictionaries-definitions-blockchain-miner. Accessed 1 May 2021
85. Parchomenko A, Nelen D, Gillabel J, Rechberger H (2019) Measuring the circular economy—A multiple correspondence analysis of 63 metrics. J Clean Prod 2019(210):200–216
86. People's Republic of China Circular Economy Promotion Law (n.d.)
87. Pomponi F, Moncaster A (2016) Circular Economy for the built environment: a research framework. J Clean Prod 143:710–718
88. Prendeville S, Cherim E, Bocken N (2017) Circular cities: mapping six cities in transition. Environ Innov Soc Trans 26:171–194
89. Reducing waste and adopting a circular economy approach will benefit the environment and create economic opportunities (n.d.). https://www.towardszerowaste.gov.sg/circular-economy/. Accessed 1 May 2021
90. Risius M, Spohrer K (2017) A blockchain research framework—What we (don't) know, where we go from here, and how we will get there. Bus Inf Syst Eng 59(6):385–409
91. Roemers G, Zande C, Thorin T, Haisma R (2018) Monitoring Voor Een Circulaire Metropool-regio. Metabolic, https://www.metabolic.nl/projects/monitoring-circularity-in-the-metropolitan-region-amsterdam/. Accessed 1 May 2021
92. Russell M, Winter J, Eijk F (2019) Circular cities, holland circular hotspot. Circle Econ. https://hollandcircularhotspot.nl/wp-content/uploads/2019/04/HCH-Brochure-20190410-web_DEF.pdf. Accessed 1 May 2021
93. Saidani M, Yannou B, Leroy Y, Cluzel F (2017) How to assess product performance in the circular economy? Proposed requirements for the design of a circularity measurement framework. Recycling 2017(2):6
94. Schmitz J, Leoni G (2019) Accounting and auditing at the time of blockchain technology: a research agenda. Aust. Account Rev 29(2):331e342. https://doi.org/10.1111/auar.12286
95. Singapore Budget (2019) Budget measures 2019. https://www.singaporebudget.gov.sg/budget_2019/budget-measures
96. Singapore Ministry of National Development Board (2017) Singapore: the first city in nature. Center for Liveable Cities
97. Singapore National Environment Agency (2019) Waste management overview. https://www.nea.gov.sg/our-services/waste-management/overview. Accessed 1 Apr 2021
98. Singapore Statutes Online (2019) Resource sustainability bill. https://sso.agc.gov.sg/BillsSupp/20-2019/Published/20190805?DocDate=20190805. Accessed 30 Mar 2021
99. Soberón M, Sánchez-Chaparro T, Urquijo J, Pereira D (2020) Introducing an organizational perspective in SDG implementation in the public sector in spain: the case of the former ministry of agriculture, fisheries, food and environment. Sustainability 12:9959
100. Soleri P, Davis M, Koolhaas R, Gehry F, Pamuk O (2013) Where we live, new perspectives quarterly, pp 13–18

101. Stark J (2016) Making sense of blockchain smart contracts. Coindesk.com. https://www.coindesk.com/making-sense-smart-contracts. Accessed 1 May 2021
102. Stockholm Resilience Center (2019) The nine planetary boundaries. https://www.stockholmresilience.org/research/planetary-boundaries/planetary-boundaries/about-the-research/the-nine-planetary-boundaries.html. Accessed 1 May 2021
103. Stroud F (2015) Blockchain: webopedia definition. https://www.webopedia.com/definitions/blockchain/. Accessed 1 May 2021
104. Suárez-Eiroa B, Fernández E, Méndez-Martínez G (2019) Operational principles of circular economy for sustainable development: linking theory and practice. J Clean Prod 214(1):952–961
105. Sultan K, Ruhi U, Lakhani R (2018) Conceptualizing blockchains: characteristics & applications. In: 11th IADIS international conference information systems 2018. University of Ottawa, Canada
106. Swan M (2015) Blockchain thinking: the brain as a decentralized autonomous corporation [Commentary]. IEEE Technol Soc Mag. https://doi.org/10.1109/MTS.2015.2494358
107. Swilling M, Hajer M, Baynes T, Bergesen J, Labbé F, Musango JK, Ramaswami A, Robinson B, Salat S, Suh S, Currie P, Fang A, Hanson A, Kruit K, Reiner M, Smit S, Tabory S (2018). The weight of cities: resource requirements of future urbanization. In: International resource panel, Nairobi. United Nations Environment Programme, Kenya. https://www.resourcepanel.org/reports/weight-cities. Accessed 2 Apr 2021
108. Sylva K (2018) A circular economic model for a sustainable city in South Asia in Sustainable development research in the asia-pacific region. In: Filho WL, Rogers J, Iyer-Raniga U (eds). Springer International Publishing
109. Tan D (2019) NEA seeks to close the waste loop and adopt a circular economy approach to resource conservation. National Environment Agency. https://www.nea.gov.sg/media/readers-letters/index/nea-seeks-to-close-the-waste-loop-and-adopt-a-circular-economy-approach-to-resource-conservation. Accessed 3 March 2021
110. Towards Zero Waste (2021) Circular economy. https://www.towardszerowaste.gov.sg/circular-economy/. Accessed 20 Mar 2021
111. UN Environment Programme (2013). Annual report 2013. http://www.unep.org/annualreport/2013/docs/ar_low_res.pdf. Accessed 20 Mar 2021
112. Upadhyay A, Mukhuty S, Kumar V, Kazancoglu Y (2021). Blockchain technology and the circular economy: implications for sustainability and social responsibility. J Clean Prod 293(2021):126130
113. Vroman HW (1994) Reinventing government: how the entrepreneurial spirit is transforming the public sector reinventing government: how the entrepreneurial spirit is transforming the public sector, vol 8. Penguin Books, New York, NY, USA
114. Weforum (2015) WEF global competitive report 2014–2015. http://www3.weforum.org/docs/WEF_GlobalCompetitivenessReport_2014-15.pdf. Accessed 20 Mar 2021
115. Wilts H (2017) Key challenges for transformations towards a circular economy—The status quo in Germany, Wuppertal Institute for climate, environment, and energy. Germany, Int Jour Waste Resour. https://doi.org/10.4172/2252-5211.1000262
116. World Bank (2020) Urban development. https://www.worldbank.org/en/topic/urbandevelopment/overview. Accessed 3 Mar 2021
117. Wörner D, Von Bomhard T, Schreier Y-P, Bilgeri D (2016) The bitcoin ecosystem: Disruption beyond financial services? In: Proceedings of 24th European conference on information systems, Istanbul, Turkey, pp 1–16
118. Zamfir V (2015) Introducing casper "the Friendly Ghost". Ethereum Fuondation Blog. https://blog.ethereum.org/2015/08/01/introducing-casper-friendly-ghost/. Accessed 1 May 2021

Leveraging Blockchain Technology in Sustainable Supply Chain Management and Logistics

Bharat Bhushan, Kaustubh Kadam, Rajnish Parashar, Shubham Kumar, and Amit Kumar Thakur

Abstract Traditional supply chain management (SCM) has some drawbacks concerning real-time and secure data about goods in transit. These require transparency to establish trust among various entities involved. Currently, it is being managed by third-party centralized controllers which make data more vulnerable. These disadvantages can be overcome by the adoption of blockchain in SCM. Blockchain is a decentralized ledger for storing, recording, managing, and transmitting data in a peer-to-peer (P2P) network. In such protocols, immutable cryptographic signatures named as hash are used to record transactions. It records information in such a manner that makes it completely impossible to alter, change, cheat or hack the system. Verification of transactions and consensus algorithms maintain data integrity and security. This paper works toward providing a brief survey on recent advances in SCM and comprehensively shows different ways of integrating blockchain technology in SCM according to different use cases. This paper comprehensively presents the evolution of blockchain technology and its current state. It provides a summary of different types of blockchain and the internal architecture of a standard blockchain protocol. The paper encapsulates some standard consensus methods such as proof of stake and proof of work. Further, the work summarizes various challenges associated with inculcating blockchain and highlights the major future research scopes in this context.

Keywords Supply chain management · Blockchain · Decentralization · IoT · Peer to peer · Logistics · Smart contracts

1 Introduction

SCM is the process of enhancing the entire business architecture by making it more flexible and agile. The main purpose of SCM is to upgrade the goods or services and to focus on how to maximize the total value and capital of the organization by efficiently

B. Bhushan (✉) · K. Kadam · R. Parashar · S. Kumar · A. K. Thakur
Department of Computer Science and Engineering, School of Engineering and Technology, Sharda University, Uttar Pradesh, India

S. S. Muthu (ed.), *Blockchain Technologies for Sustainability*,
Environmental Footprints and Eco-design of Products and Processes,
https://doi.org/10.1007/978-981-16-6301-7_9

managing the resources. SCM represents the attempts by the producers to create and implement the supply chains in a more feasible and effective way aiming to centrally control the production of the goods. There are five basic elements in an SCM system that are planning, sourcing, making, delivering, and returning. Blockchain is one of the most evolving technologies in the present world and has a versatile implementation for trade arrangements, cross-border transactions, business decentralization, and many more. Blockchain is a decentralized system of managing data and information implemented on a P2P network [1]. Blockchain, being decentralized, is capable of eliminating a single point of failure by using immutable documentation of the data and authorized admittance of the user [2]. Transparency, traceability, security, efficiency, confidentiality, and immutability are the key features that can be ensured in the SCM system with the implementation of blockchain technology [3].

Various challenges in the supply chain can be solved based on the above-mentioned blockchain features. There is a wide array of use cases of blockchain in SCM like invoice processing, logistics, product traceability, etc. Blockchain helps industries like E-commerce, logistics, agriculture, food, etc. [4]. The tokenization of a product or an asset helps to make it tradeable [5]. Blockchain when added with the Internet of Things (IoT) devices can be used for traceability in the food and pharmaceutical industries [6]. IoT devices can record different metrics such as temperature, humidity, and vibration [7]. In order to ensure automatic redress, blockchain and smart contracts are implemented in case the readings fluctuate out of the range [8]. The implementation of the blockchain using smart contracts that can automatically redress the errors in the data displays a splendid example of the recent advances in SCM and how it can be decentralized and can be made more efficient by using blockchain. A summary of the contributions of this work is mentioned below.

- This work presents an introduction about the SCM and different use cases of blockchain in SCM and how it can be implemented using different mechanisms.
- This work presents an in-depth survey and explores state-of-the-art blockchain technology highlighting its emergence, classification, characteristics, architecture, and consensus mechanisms.
- This work demonstrates the processes involved in the working of blockchain.
- This work highlights the need for leveraging blockchain technology in SCM and also discusses different use cases of blockchain in SCM.

The remainder of the paper is organized as follows. Section 2 is a literature review that explores all the previous research efforts made and different ways to solve the problems of logistics and supply chain. Section 3 extends an outline of blockchain technology and gives a summary of consensus algorithms; it also discusses blockchain transactions and the steps involved to execute them. Section 4 throws light on the need for integration of blockchain and various ways to implement blockchain to improve SCM. It also highlights different use cases where blockchain can be used in SCM. Section 5 contains the inference and future research directions in the field.

2 Literature Review

Madhumita et al. [9] developed a very effective protocol that is based on a lightweight blockchain that uses radio frequency identification (RFID)-based authentication. This system is developed for a 5G mobile computing environment called lightweight blockchain-enabled radio frequency identification (LBRAPS). LBRAPS uses operations like one-way cryptographic hash, bitwise exclusive-or (XOR), bitwise rotation only. This mobile computing environment is secured against attacks. The trade-off in functionality, communication, computation costs, and features of LBRAPS is better than that of existing protocols. Chang et al. [10] study comprehends and provides a design to apply blockchain in the management of the supply chain from a point of view of literary analysis. It also highlights future efforts regarding supply chain and blockchain integration and their social impact. Malik et al. [11] propose a three-layer framework called Trustchain that tracks interaction among participants of the supply chain. It actively assigns scores considering all the interactions and reputation. The novelty of Trustchain stems forms a model that evaluates reputation on parameters like entity trustworthiness, events on the blockchain, quality of commodities, also its support for reputation scores that separate between products and a supply chain participant, enabling the allocation of product-specific reputations for the same participant and uses smart contracts for automated and unbiased calculation of scores.

Rejeb et al. [12] derive research propositions of how features of IoT like immutability, traceability, quality, security, etc., can be impacted by blockchain. It also illustrates how the integration of blockchain and IoT can help applications in the modern supply chain. Sahai et al. [13] propose a solution that makes cryptographic accumulators zero knowledge proofs that guarantee privacy as well as traceability. They provide an evaluation for implementing their protocol on hyperledger fabric. Their work illustrates a blockchain-based model for the supply chain that gives an efficient tracing for contamination while accounting for traceability as well as privacy. Shakhbulatov et al. [14] provide a new way to track carbon footprint on food production and its various supply chain stages. They design a system to do so by using a cluster-based record keeping the private data safe. The system implementation is evaluated on its latency under various conditions. They also show blockchain implementation is scalable with a large number of rows. Niya et al. [15] demonstrate the implementation of a supply chain application that works on the Ethereum blockchain. This decentralized application (DApp) gives a general-purpose platform as well as a hardware-independent approach that facilitates object combinations and tracking of transformation.

Wu et al. [16] give a detailed analysis of new requirements, potential opportunities, and provide methodologies for developing blockchain-based SCM systems. They also provide solutions for technical challenges in terms of access control, scalability, and data retrieval. They have also done a case study for designing a system for food traceability and provide more insights for technical challenges. Hinckeldeyn et al. [17] present a prototype that implements smart contracts using smart storage containers to study the potential of IoT and blockchain in logistics. They present

Table 1 Comparative summary of existing related surveys

Reference	Year	Contribution
Madhumita et al. [9]	2019	Used RFID-based proof mechanism to design a blockchain system for a mobile computing environment making the system reliable against the attacks
Chang et al. [10]	2020	Worked on the literary survey to provide a model to implement blockchain in SCM
Malik et al. [11]	2019	Developed Trustchain, a three-layered architecture model to track the exchange between supply chain members
Rejeb et al. [12]	2019	Worked on the integration of the blockchain with IoT for the purpose of enhancing the supply chain
Sahai et al. [13]	2020	Proposed a method based on the cryptographic accumulator's zero knowledge proof aiming to enhance privacy and improve tracing
Shakhbulatov et al. [14]	2019	Designed a system for carbon footprint tracking on foods using the collection of data implementing blockchain
Niya et al. [15]	2019	Proposed the supply chain based on the Ethereum blockchain for tracking and decentralization of products
Wu et al. [16]	2019	Presented a case study to deal with technical challenges in the supply chain and enhance food traceability
Hinckeldeyn et al. [17]	2018	Developed a smart storage container for a blockchain-based supply chain using Ethereum-based smart contracts
Wen et al. [18]	2019	Introduced a mechanism to align IIoT to the blockchain and proposed a fine-graded data sharing method for the supply chain

a system where an Ethereum-based smart contract is connected to a smart storage container. The proposed smart contract is developed to process payment through a multi-signature wallet of three parties. They also provide further research implications from this prototype study. Wen et al. [18] propose a system to connect Industrial Internet of Things (IIoT) devices that record and monitor real-time data in the network to the blockchain by implementing smart contracts. The blockchain-based supply chain structure provides collaborative solutions between all entities. This structure uses a fine-grained data sharing scheme. The smart contract can only be accessed and executed by companies and entities that satisfy access policies. This scheme ensures the reliability of data and privacy. A comparative study of existing work in this area of study is summarized in Table 1.

3 Basics of Blockchain

The blockchain concepts were first proposed in 2008. It is a system or a specific type of decentralized database which records data in such a way that makes it completely impractical to alter, change, cheat, and hack the system. It is a kind of digital ledger that keeps transactions in a form of records that are further distributed and duplicated across the entire blockchain network. Each block in the chain consists of some data and is chained together along with the number of transactions. The blockchain maintains a chronological order, whenever a new data comes into a new block, it gets chained with the previous block and a record of that transaction is added to the ledger of every participant. Blockchain works on distributed ledger Technology (DLT) in which an immutable cryptographic signature named as the hash is used to record transactions.

3.1 History of Blockchain

The first blockchain technology was introduced by W. Scott Stornetta and Stuart Haber in 1991. They proposed a solution for the security of digital documents, and these documents can be secured using a time-stamping technique so that they could not be tampered with and backdated. For this, they developed a secured chain of blocks using cryptography to store the time-stamped documents. In 1992, the concept of Merkle trees was introduced, which makes a more efficient blockchain because using it one can store multiple documents in one block [19]. A secured chain of blocks can be created using Merkle trees. Record of each data is connected to the previous one before it along with the series of data records stored in the block. The patent for this lapsed in 2004 and this technology went unused.

In 2004, a system called reusable proof of work (RPoW) was developed as a prototype for digital cash by cryptographic activist and computer scientist Hal Finney. He developed an RSA-signed token that can be further transferred from one person to another. In 2008, Satoshi Nakamoto introduced the concept of distributed blockchains. According to his introduced design, appending a block in the chain does not require any sign by trusted parties. He used the P2P network for verifying and time-stamping each transaction. These modifications were so efficient that makes blockchain very useful in the modern-day cryptocurrency space. Different stages in the evolution of blockchain technology are summarized in Table 2.

3.2 Architecture of Blockchain

A blockchain is designed as a completely decentralized network, and its architecture maintains a list that can update the ordered records, commonly referred to as blocks. As described in Fig. 1, each block maintains a time stamp, current and previous block's hash, and data inside the block. A blockchain network has the following architectural components:

Table 2 History of blockchain

Year	Brief history
1991	W Scott and Stuart Haber Stornetta described a chain of blocks that is cryptographically secured for the first time
1998	Bit gold, a digital currency, was introduced by computer scientist Nick Szabo
2000	Stefan Konst proposed ideas for the implementation of cryptographically secured chains and also published his theory
2004	A prototype for digital cash was developed by Hal Finney based on a system called RPoW
2008	Model for a blockchain introduced on a white paper by developers working under Satoshi Nakamoto
2009	First blockchain technology was implemented by Nakamoto for bitcoin's public ledger for transactions
2014	Blockchain 2.0 was born and separated from the currency. Blockchain technology is explored for other domains like inter-organizational, financial transactions

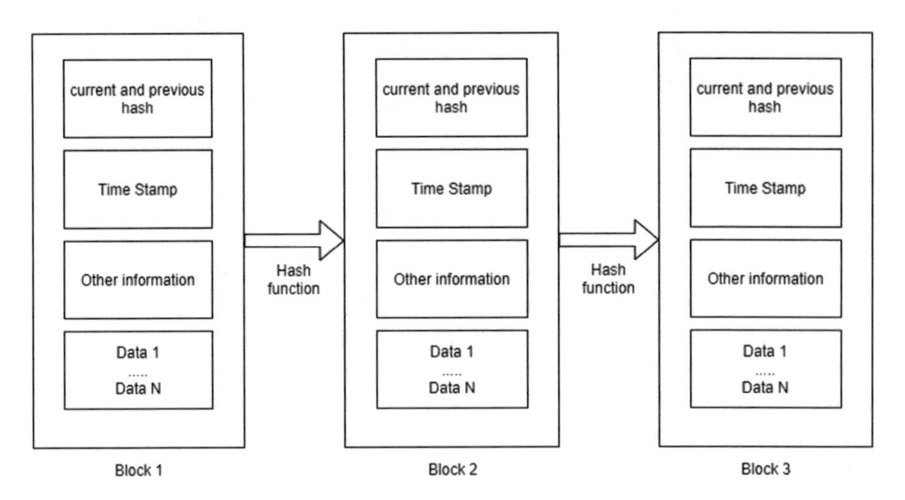

Fig. 1 Architecture of blocks

3.2.1 Transaction

Transactions can be termed as basic building blocks of a blockchain system. Mainly they have a sender address, a recipient address, and a value. The value is digitally signed by the hash generated through the public key of the receiver and adding the previous transaction. These transactions are bundled together in a form of blocks and then delivered to each node in the network. After the distribution of transactions throughout the network, they are independently processed and verified by each node, and then each transaction is time-stamped and collected in a block.

3.2.2 Block

Information like transactions and block headers resides in a block. Blocks are basically a kind of data structure used to bundle sets of transactions and then added to all nodes in the network. Blocks are generated through a process known as mining. Mining is used to create a valid and acceptable block for the entire network [20]. There are mainly three types of blocks: blocks of the main branch, blocks of the side branch, and orphan blocks [21].

- Main branch blocks are used to add to the main blockchain.
- Side branch blocks are used as references to the parent block.
- Orphan blocks are used as references to parent blocks that are unknown to the node processing the block.

3.2.3 P2P Network

Blockchain uses P2P networks to make them decentralized and more secure. Every participant of the network will get the full data of the blockchain [22]. This P2P network in blockchain works on the Internet protocol which has a flat topology with no centralized node. A P2P network is not like a centralized network and does not have a single point of failure or attack that is why they are more secure. These peers in the network are used as storage and computing power that is required for the maintenance of the network.

3.2.4 Consensus Algorithm

Consensus algorithms are used to maintain the copies of a single ledger in a synchronized way. The nodes in the network use these algorithms to make their local copies consistent and updated [23]. Firstly, a consensus algorithm sets up a process to verify, validate, and confirm transactions and then records those transactions in a large distributed directory, creation of a block record, i.e., blocks in a form of a chain, and finally implementation of consensus protocol. Thus, verification, validation, immutable recording, and consensus lead to the security and trust of the blockchain. Some of the consensus algorithms widely used in blockchain are as follows:

- **Proof of Work (PoW)**: PoW tackles a computationally challenging task while creating a new block in the blockchain network. It is also used for the selection of a miner for the generation of the next block. It is basically used for guessing the string that generates a 256-bit hash, produced by the SHA256 algorithm [24]. Solving the complex mathematical problem is called mining, and in cryptocurrency miners are usually rewarded for their work. The odds of solving these problems are about 1 in 5.9 trillion, and trial and error is the only method used to

Table 3 Comparison of various consensus protocols

	Speed	Consumption of energy	Security	Centralization degree
PoW	Slow	Very high	Secure	Very low
PoS	Normal	Normal	Secure	Low
PoB	Normal	Very high	Secure	Very high
PoET	Normal	Low	Secure	Very low

solve these. A considerable amount of energy in the form of substantial computing power is required in order to solve these complex mathematical problems.

- **Proof of Stake (PoS)**: PoS validates the transactions to get transaction fees. It was introduced as an alternative to PoW. In this, nodes work as validators and are randomly assigned to validate. The probability of this selection depends on the particular node's stake. For the miner, PoS makes an attack less advantageous by structuring compensation. Currently, the concept of PoS is only used by altcoins. When it comes to unrelated areas such as logistics, big data, artificial intelligence, and other mathematical fields, then PoS is more favored than PoW.
- **Proof of Burn (PoB)**: PoB is used as a transaction tool in the blockchain that improves the functioning of the blockchain. It is used as a more sustainable alternative consensus to resolve the high-energy consumption problem of PoW that is why it is often known as PoW without energy waste. It does not depend on mining hardware and does not require powerful computational resources. Some of the advantages of PoB are more sustainability, no need for mining hardware, less centralized coin distribution or mining, and reduced circulating supply with the help of coin burns.
- **Proof of Elapsed Time (PoET)**: PoET was developed by Intel Corporation that determines mining rights and block winners for a permission blockchain. Basically, it works on a lottery system in which all participating nodes across the network, present in the blockchain, have an equal chance of winning. Each node must go to sleep for a random amount of wait time that is generated by the PoET. The node that wakes up first because of its short wait time will win the block. The workflow of PoET is quite similar to PoW, but PoET consumes less power because of its increased efficiency as it allows a miner's processor to switch to another task and sleep for the specified time. Table 3 shows some of the differences between various consensus algorithms used in blockchain.

3.3 Types of Blockchain

There are mainly three main types of blockchain that are public blockchain, private blockchain, and consortium blockchain. Each one of them is used to fulfill a problem or a certain set of problems. Different types of blockchains are explored in the subsections below.

3.3.1 Public Blockchain

Public blockchain works on the concept of distributed ledger technology (DLT). This kind of blockchain is mainly used in cryptocurrencies like Bitcoin [25]. It eliminates the problems that occur with centralized systems, i.e., transparency and security. DLT distributes data across a P2P network and does not store data in any one spot. Its decentralized nature requires some technique for confirming the genuineness of information. High security, an open environment, no regulations, and anonymous nature are some of the features of a public blockchain.

3.3.2 Private Blockchain

This blockchain network has a restrictive environment basically like a closed network, and only authorized participants can access it [26]. It is managed and controlled by a privately owned organization. Specific rights and restrictions are granted to the participant in this blockchain. This blockchain is not fully decentralized like a public blockchain. High efficiency, full privacy, and stability are some of the features of private blockchain.

3.3.3 Consortium Blockchain

The fourth kind of blockchain, consortium blockchain, also called a semi-decentralized type of blockchain, is like a hybrid blockchain in that it has private and public blockchain features [27]. This type of blockchain is managed by more than one organization. It is quite similar to a private blockchain, and also more like a permission platform, not a public platform. Faster speed, scalability, and low transaction costs are some of the features of consortium blockchain.

Table 4 shows the difference between all three types of blockchain taking into consideration characteristics like access, consensus mechanisms, use cases, etc.

3.4 Working of Blockchain

Since blockchain is used as a decentralized ledger for all transactions across a P2P network, using these network technology, participants do not require any central authority for the confirmation of transactions. Figure 2 shows multiple steps associated with the working of a blockchain.

As mentioned in Fig. 2 firstly, a transaction is requested by a user, and corresponding to that a secured block is created, representing that transaction is created. Each block contains three basic elements, i.e., data, a nonce, and a hash in the block.

Table 4 Different types of blockchain

Parameters	Private blockchain (requires permission)	Public blockchain (permissionless)	Consortium or hybrid blockchain
Access	Single organization	Anyone	Multiple selected organizations
Network type	Partially decentralized	Decentralized	A hybrid between private and public blockchain
Determination of consensus	Set of selected nodes	All miners in the network	Within one organization
Advantages	Controlled access, high performance, and security	Independent, transparency, and trustworthy	Scalability, security, and access control
Disadvantages	Less trustworthy and auditability	Performance, security, and scalability	Transparency
Use cases	Asset ownership, and supply chain	Cryptocurrency and document validation	Research, supply chain, and banking

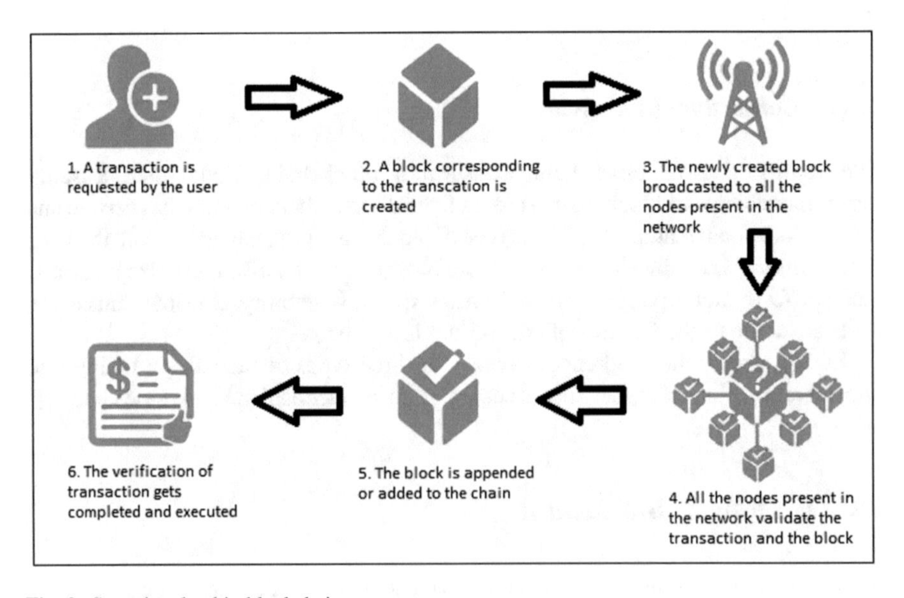

Fig. 2 Steps involved in blockchain

- A nonce is a whole number of 32-bit that is used for the generation of a block header hash [28].
- A 256-bit number associated with the nonce is known as a hash.

A nonce generates a cryptographic hash when the creation of the first block occurs. The data that resides in the block is treated as forever tied and signed with the hash and nonce unless it is further mined by the miners. Miners help in creating a new block in the chain through a process termed mining [29]. Each and every block present in the chain excluding the genesis block that is the first block in the chain contains the hash of the previous block along with its own unique hash and nonce. To generate an acceptable hash, miners use software that can solve complex problems for finding a nonce. Making any kind of change or altering any block that is present in the chain requires re-mining of all the other blocks that come after that particular block.

After successful mining or creation of a block, the requested transaction is broadcasted over the P2P network, as mentioned in step 3 in Fig. 2, to validate it by the nodes present in the network [30]. Any electronic device which has copies of the blockchain and can also maintain those copies can be termed as nodes in the blockchain. In order to append a new block or make a change in the existing block the requested transaction must be algorithmically approved by every node associated with that blockchain network. Lastly, once the nodes verify the transaction, the new block is appended to the existing blockchain as mentioned in the last step of Fig. 2. These blocks are bound and secured to each other using cryptographic algorithms.

4 Supply Chain Management

The supply chain has a big role to play in the world's economy as it accounts for more than 76% of the global trade. Tracks of the supply chain are found way back since 1911, when the founder of industrial engineering, Fredrick Taylor, tried scientific management to improve the manual loading in his work. Between 1940 and 1950, to obtain better warehouse space, layout and racking, pallet and pallet lifts mechanization was researched to improve logistics. During the 1980s, logistics transformation based on personal computing began to improve SCM. In the 1990s, enterprise resource planning (ERP) systems came into existence to boost logistics. Globalization in the manufacturing process during the mid-1990s helped spread the term "supply chain". With the increase in the industries and productions, the supply chain became a major part of their development process. Implementation of supply chain management nowadays is widely spread in the industry in one or the other form. Inventory management, warehouse management, logistics and transportation, production, customer service, security, business, e-commerce, etc., are some of the fields which are operated based on different SCM methodologies [31].

SCM includes multiple stakeholders and participants and several processes in various steps. SCM's efficiency is measured against the competitive nature of the industry to sustain and to provide quality goods and services to the end user coping up with high risk and greater demand. Sometimes it is difficult to keep records of ownership, processes, and materials at multiple steps [32]. The kind of SCM which exists today has various limitations [33]. Rapidly increasing demand to operate more transactions and services in the fast-paced world globally is tough. Traditional supply chain contract architecture is developed based on the asymmetric data of supplier reliability [34]. Ensuring transparency, flexibility, and reduction to risk factors and meeting the needs of end users can be managed efficiently using blockchain technology. SCM has become very essential in any 34th blockchain-based industry for its improvement and accuracy in real-time information about products, in the in-transit movements, and in making correct decisions for supply chain operations [35].

4.1 Motivation for Applying Blockchain in SCM

There are several drawbacks in the traditional SCM which cause problems in its implementation in the industries. One of the major problems is information asymmetry where the length of the supply chain is very large which results in spamming the entire SCM from the producers to the clients. Due to such a vast implementation of the supply chain, it needs to be more reliable than it has been earlier. Any fault in the supply chain will cause a considerable loss for the industry. In order to fulfill the dynamically changing requirements, a very reliable technology such as blockchain provides a suitable solution for the same [36]. Blockchain is capable of eliminating a single point of damage due to its decentralized behavior by using immutable documentation of the data and authorized admittance of the user.

There are several requirements that can be efficiently and effectively satisfied by the implementation of blockchain, hence there is no better choice than blockchain for industries to manage their respective supply chain. Firstly, traceability is the most important and frequent use case for management in the quality of sensitive products such as food or medicine. Secondly, provenance is required for keeping records of product ownership through the supply chain. From origin to the final stage, blockchain should grant records of a product at each stage. Thirdly, distribution is required because the transaction and the activities in the supply chain will not be managed by any centralized organization, and each node in the supply chain keeps the record of the state of the transaction, any fault at any network node would not result in the entire supply chain to be destroyed. Lastly, immutability, sustainability, data privacy, confidentiality, and transparency are some of the other major factors considering the implementation of blockchain in SCM [37].

Blockchain works as a record-keeping system. Blockchain ledgers are basically transaction records that are managed by multiple nodes which makes it impossible to manipulate the transaction records. A blockchain-based supply change can contribute to recording price, location, date, certification, quality, and many such related data to be managed more efficiently. The traceability of goods and products in the supply chain can be improved using blockchain and also helps in reducing the chance of loss in the gray market which, in turn, enhances the transparency throughout the process of production to supply and also boosts the position as a leader in fair production industries [38].

4.2 How Blockchain Can Be Implemented?

There are supply chain challenges that can be solved based on the following blockchain protocols. Hash map table, a secondary data structure can be used to operate with the data in the blockchain. Different consensus algorithms and blockchain frameworks have been proposed to deal with the problems in the products and services supply chain in specific industries. Blockchain can be implemented by Ethereum and InterPlanetary File System (IPFS) which is a decentralized mechanism with efficient and fast data tracking speeds and unchallenging access to data for a huge count of stakeholders at different levels in the supply chain [39]. In order to input data to the blockchain, standard IoT devices and RFID tags can be used on products such as food and wood industries, which can add labels to the products and the same can be stored in the corresponding blockchain [40].

Smart contracts can be of potential use for blockchain implementation in SCM. In order to avoid managerial risks, thievery, fraud, or other miseries, agreements and regulated conditions can be written in computer code for implementing smart contracts, and when the blockchain network gets added up by these smart contracts, they are shared to all the linked nodes [41]. Any new change in a particular database can activate the given conditions which lead to the execution of related process flow. Smart contracts may help in the automation of the process and execution of the obligation as well. There is a wide array of use cases of blockchain in SCM like invoice processing, logistics, product traceability, etc. Blockchain helps the e-commerce industry by increasing the efficiency in invoice and payment processing [42]. Moreover, payments can be automated by using smart contracts. This is a faster process than the traditional process that may take up to thirty days. The tokenization of a product or an asset helps to make it tradeable. Hence, the end user can assign ownership without involving any physical transfer of assets. Blockchain allied with IoT devices can be used for traceability in the food and pharmaceutical industries. IoT devices can record different metrics such as temperature, humidity, and vibration. The information is kept in a blockchain, and smart contracts are implemented to make sure the self-regulating correctness if any of the feeds go out of the defined range.

Many food safety concerns, such as contamination, can be tracked and isolated. Many issues arise due to less visibility in the supply chain and slow responses to them [43]. A consortium, in order to make sure the transparency of the entire product transportation and its position, is utilizing blockchain inside the supply chain [44]. Modern supply chains are handled by third parties, and therefore, the interaction between suppliers, providers, and customers is minimized. The logistics industry can be reinvented by the use of blockchain as it provides verifiability to transactions and can be coordinated autonomously without involving a third party. Blockchain helps an individual to authenticate the source of the product as well as its transportation from the producer to the end users. Blockchain in SCM can also revolutionize the energy industry. It may enable users to sell excess power to the distributor. Figure 3 shows the implementation of blockchain in SCM using various methodologies.

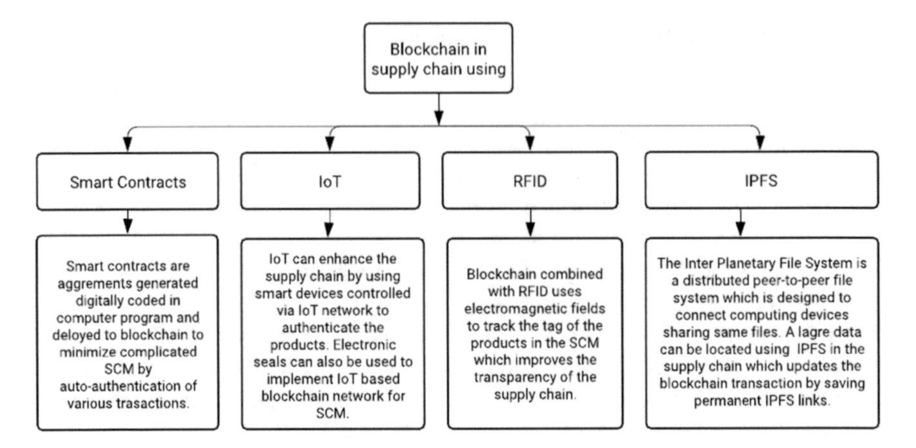

Fig. 3 Implementation of blockchain in SCM

4.3 Opportunities for Integration of Blockchain and SCM

Blockchain provides increased transparency and efficient and cost-effective scalability with reduced risk across SCM. Some of the chief potential benefits of applying blockchain in SCM are increased visibility over contract manufacturing, meeting corporate standards by increasing traceability of material supply, reduced counterfeit products, and lower administrative costs. Some of the secondary potential benefits of applying blockchain in SCM are to reduce potential public relations risk from malpractices of the supply chain, to provide transparency of materials used in products that lead to strengthening the reputation of corporate, to engage stakeholders, and to improve the public trust and credibility of data shared. Listed below are some potential use cases of blockchain in SCM and descriptions of crucial solutions to them.

4.3.1 Provenance for Food and Beverage

The food sector always tackles frauds, scandals, and inefficiencies of the supply chain. Today our food system does not meet the assurance and transparency demanded by many customers. This problem can be eliminated by a collaboration of participants over a blockchain network which can ensure transparency and a trusted data platform. Consumers can know detailed information about the product such as the geographic location, logistics information, source and ingredients of the products, inspection report, even temperature data by the use of quick response (QR) code, and the party who produces data in the blockchain will time-stamp and cryptographically sign the data accordingly [45].

4.3.2 Digitization for High-Value Products and Anti-Counterfeiting

The amount of counterfeiting has reached 1.82 trillion dollars USD by 2020, which has increased concern over the legitimacy of pre-owned luxury products in secondary markets. Blockchain allows the digitization of products and establishes a link between physical commodity and unique identity on the blockchain through the use of near field communication (NFC) tags. With the use of this digitized identity, we can track the life cycle of products from manufacturing to consumer. Moreover, ownership of the product can be tracked by tying it to the account of the user and can be transferred in the case of secondary markets [46].

4.3.3 Digital Asset Passport

Using blockchain technology frauds in asset exchange can be eliminated. The authorized parties can maintain and create the digital asset passport by attaching a digital ledger along with registering it on the blockchain. Ownership data of the asset will be stored, secured, and time-stamped by the blockchain. Read and write access to the digital asset can be controlled by the asset owners, and potential buyers can see the information from the collected data. When the asset is sold, the data associated with that particular asset will be added and replicated in the blockchain accordingly.

4.3.4 Digital Low-Carbon Emission Ecosystem

Various concerns about global warming, carbon emissions, and pollution have been raised by international organizations. Some of the issues that hinder current progress in lowering carbon emissions are non-quantifiable nature, unfair execution of policies, and lack of effective regulation. Implementation of blockchain can digitize and quantify user behavior. Validation of data of user's behavior can be done through the execution of smart contracts along with transparent proofs [47]. User's behavior can be marked by IoTs and the others from different use cases such as electric charging station use, driving electric vehicles, public transportation use, carpooling, and eco-friendly household appliances.

5 Conclusion

This article addresses an important problem of integrating blockchain in SCM. Orthodox SCM has various weaknesses regarding data integrity, security, and traceability. Blockchain provides automation, and all the drawbacks of SCM are overcome by integrating it. Decentralized systems empower permissionless ownership. Consensus is supposed to make a decision that means a group of nodes approve the transactions as opposed to an individual node. Distributed ledger on P2P networks provides a way to store data and makes it immutable. Blockchain also provides tokenization of goods to be transported. The paper summarizes and explores different ways of adapting blockchain into traditional SCM. This paper also shows various ways to implement blockchain depending on various use cases in SCM. It also discusses various trade-offs associated with blockchain and SCM. It also highlights directions where advances can be made in future concerning blockchain and SCM.

References

1. Ramaswamy L, Gedik B, Liu L (2005) A distributed approach to node clustering in decentralized peer-to-peer networks. IEEE Trans Parall Distrib Syst 16(9):814–829.https://doi.org/10.1109/tpds.2005.101
2. Haque AK, Bhushan B, Dhiman G (2021) Conceptualizing smart city applications: requirements, architecture, security issues, and emerging trends. Expert Syst. https://doi.org/10.1111/exsy.12753
3. Bhushan B, Sahoo C, Sinha P, Khamparia A (2020) Unification of blockchain and internet of things (BIoT): requirements, working model, challenges and future directions. Wirel Netw. https://doi.org/10.1007/s11276-020-02445-6
4. Yang X, Li M, Yu H, Wang M, Xu D, Sun C (2021) A trusted blockchain-based traceability system for fruit and vegetable agricultural products. IEEE Access 9:36282–36293. https://doi.org/10.1109/access.2021.3062845
5. Saxena S, Bhushan B, Ahad MA (2021) Blockchain based solutions to secure IoT: background, integration trends and a way forward. J Netw Comput Appl 103050. https://doi.org/10.1016/j.jnca.2021.103050
6. Sun S, Du R, Chen S, Li W (2021) Blockchain-based IoT access control system: towards security, lightweight, and cross-domain. IEEE Access 9:36868–36878. https://doi.org/10.1109/access.2021.3059863
7. Misra NN, Dixit Y, Al-Mallahi A, Bhullar MS, Upadhyay R, Martynenko A (2020) IoT, big data and artificial intelligence in agriculture and food industry. IEEE Internet of Things J 1–1. https://doi.org/10.1109/jiot.2020.2998584
8. Xu J, Wang S, Zhou A, Yang F (2020) Edgence: a blockchain-enabled edge-computing platform for intelligent IoT-based dApps. China Commun 17(4):78–87. https://doi.org/10.23919/jcc.2020.04.008
9. Madumidha S, Ranjani PS, Varsinee SS, Sundari PS (2019) Transparency and traceability: in food supply chain system using blockchain technology with internet of things. In: 2019 3rd international conference on trends in electronics and informatics (ICOEI). https://doi.org/10.1109/icoei.2019.8862726
10. Chang SE, Chen Y (2020) When blockchain meets supply chain: a systematic literature review on current development and potential applications. IEEE Access 8:62478–62494. https://doi.org/10.1109/access.2020.2983601
11. Malik S, Dedeoglu V, Kanhere SS, Jurdak R (2019) TrustChain: trust management in blockchain and IoT supported supply chains. In: 2019 IEEE international conference on blockchain (blockchain). https://doi.org/10.1109/blockchain.2019.00032
12. Rejeb A, Keogh JG, Treiblmaier H (2019) Leveraging the internet of things and blockchain technology in supply chain management. Future Internet 11(7):161. https://doi.org/10.3390/fi11070161
13. Sahai S, Singh N, Dayama P (2020) Enabling privacy and traceability in supply chains using blockchain and zero knowledge proofs. In: 2020 IEEE international conference on blockchain (blockchain). https://doi.org/10.1109/blockchain50366.2020.00024

14. Shakhbulatov D, Arora A, Dong Z, Rojas-Cessa R (2019) Blockchain Implementation for analysis of carbon footprint across food supply chain. In: 2019 IEEE international conference on blockchain (blockchain). https://doi.org/10.1109/blockchain.2019.00079
15. Niya SR, Dordevic D, Nabi AG, Mann T, Stiller B (2019) A platform-independent, generic-purpose, and blockchain-based supply chain tracking. In: 2019 IEEE international conference on blockchain and cryptocurrency (ICBC). https://doi.org/10.1109/bloc.2019.8751415
16. Wu H, Cao J, Yang Y, Tung CL, Jiang S, Tang B, Liu Y, Wang X, Deng Y (2019) Data management in supply chain using blockchain: challenges and a case study. In: 2019 28th international conference on computer communication and networks (ICCCN). https://doi.org/10.1109/icccn.2019.8846964
17. Hinckeldeyn J, Jochen K (2018) (Short Paper) developing a smart storage container for a blockchain-based supply chain application. In: 2018 crypto valley conference on blockchain technology (CVCBT). https://doi.org/10.1109/cvcbt.2018.00017
18. Wen Q, Gao Y, Chen Z, Wu D (2019) A blockchain-based data sharing scheme in the supply chain by IIoT. In: 2019 IEEE international conference on industrial cyber physical systems (ICPS). https://doi.org/10.1109/icphys.2019.8780161
19. Dhumwad S, Sukhadeve M, Naik C, Manjunath KN, Prabhu S (2017) A peer to peer money transfer using SHA256 and Merkle tree. In: 2017 23RD annual international conference in advanced computing and communications (ADCOM). https://doi.org/10.1109/adcom.2017.00013
20. Bhushan B, Sinha P, Sagayam KM, Andrew A (2021) Untangling blockchain technology: a survey on state of the art, security threats, privacy services, applications and future research directions. Comput Electr Eng 90:106897. https://doi.org/10.1016/j.compeleceng.2020.106897
21. Viriyasitavat W, Da Xu L, Bi Z, Sapsomboon A (2019) New blockchain-based architecture for service interoperations in the internet of things. IEEE Trans Comput Soc Syst 6(4):739–748. https://doi.org/10.1109/tcss.2019.2924442
22. Hao W, Zeng J, Dai X, Xiao J, Hua Q-S, Chen H, Li K-C, Jin H (2020) Towards a trust-enhanced blockchain P2P topology for enabling fast and reliable broadcast. IEEE Trans Netw Serv Manag 17(2):904–917. https://doi.org/10.1109/tnsm.2020.2980303
23. Zhou T, Li X, Zhao H (2019) DLattice: a permission-less blockchain based on DPoS-BA-DAG consensus for data tokenization. IEEE Access 7:39273–39287. https://doi.org/10.1109/access.2019.2906637
24. Tran TH, Pham HL, Nakashima Y (2021) A high-performance multimem SHA-256 accelerator for society 5.0. IEEE Access 9:39182–39192. https://doi.org/10.1109/access.2021.3063485
25. Aleksieva V, Valchanov H, Huliyan A (2020) Smart contracts based on private and public blockchains for the purpose of insurance services. In: 2020 international conference automatics and informatics (ICAI). https://doi.org/10.1109/icai50593.2020.9311371
26. Baucas MJ, Gadsden SA, Spachos P (2021) IoT-based smart home device monitor using private blockchain technology and localization. IEEE Netw Lett 3(2):52–55. https://doi.org/10.1109/lnet.2021.3070270
27. Cui Z, Xue F, Zhang S, Cai X, Cao Y, Zhang W, Chen J (2020) A hybrid blockchain-based identity authentication scheme for multi-WSN. IEEE Trans Serv Comput 1–1. https://doi.org/10.1109/tsc.2020.2964537
28. Kamal M, Tariq M (2019) Light-weight security and blockchain-based provenance for advanced metering infrastructure. IEEE Access 7:87345–87356. https://doi.org/10.1109/access.2019.2925787
29. Chen R, Tu I-P, Chuang K-E, Lin Q-X, Liao S-W, Liao W (2020) Index: degree of mining power decentralization for proof-of-work based blockchain systems. IEEE Netw 34(6):266–271. https://doi.org/10.1109/mnet.011.2000178
30. Liang Y, Lu C, Zhao Y, Sun C (2021) Interference-based consensus and transaction validation mechanisms for blockchain-based spectrum management. IEEE Access 9:90757–90766. https://doi.org/10.1109/access.2021.3091802

31. Habib M (2011) Supply chain management (SCM): theory and evolution. In: Supply chain management—Applications and simulations. InTech. https://doi.org/10.5772/24573
32. Omar IA, Jayaraman R, Salah K, Debe M, Omar M (2020) Enhancing vendor managed inventory supply chain operations using blockchain smart contracts. IEEE Access 8:182704–182719. https://doi.org/10.1109/access.2020.3028031
33. Du M, Chen Q, Xiao J, Yang H, Ma X (2020) Supply chain finance innovation using blockchain. IEEE Trans Eng Manag 67(4):1045–1058. https://doi.org/10.1109/tem.2020.2971858
34. Bhushan B, Khamparia A, Sagayam KM, Sharma SK, Ahad MA, Debnath NC (2020) Blockchain for smart cities: a review of architectures, integration trends and future research directions. Sustain Cities Soc61:102360. https://doi.org/10.1016/j.scs.2020.102360
35. Gaonkar RS, Viswanadham N (2007) Analytical framework for the management of risk in supply chains. IEEE Trans Autom Sci Eng 4(2):265–273. https://doi.org/10.1109/tase.2006.880540
36. Song JM, Sung J, Park T (2019) Applications of blockchain to improve supply chain traceability. Procedia Comput Sci 162:119–122. https://doi.org/10.1016/j.procs.2019.11.266
37. Gonczol P, Katsikouli P, Herskind L, Dragoni N (2020) Blockchain implementations and use cases for supply chains-a survey. IEEE Access 8:11856–11871. https://doi.org/10.1109/access.2020.2964880
38. Zheng W, Zheng Z, Chen X, Dai K, Li P, Chen R (2019) NutBaaS: a blockchain-as-a-service platform. IEEE Access 7:134422–134433. https://doi.org/10.1109/access.2019.2941905
39. Battah AA, Madine MM, Alzaabi H, Yaqoob I, Salah K, Jayaraman R (2020) Blockchain-based multi-party authorization for accessing IPFS encrypted data. IEEE Access 8:196813–196825. https://doi.org/10.1109/access.2020.3034260
40. Wang S, Zhu S, Zhang Y (2018) Blockchain-based mutual authentication security protocol for distributed RFID systems. In: 2018 IEEE symposium on computers and communications (ISCC). https://doi.org/10.1109/iscc.2018.8538567
41. Al-Jaroodi J, Mohamed N (2019) Blockchain in industries: a survey. IEEE Access 7:36500–36515. https://doi.org/10.1109/access.2019.2903554
42. Daoyuan S (2011) Research on supply chain management in enterprises based on E-commerce. In: 2011 international conference on computer science and service system (CSSS). https://doi.org/10.1109/csss.2011.5974828
43. Zhu Q, Kouhizadeh M (2019) Blockchain technology, supply chain information, and strategic product deletion management. IEEE Eng Manag Rev 47(1):36–44. https://doi.org/10.1109/emr.2019.2898178
44. Zeng X, Hao N, Zheng J, Xu X (2019) A consortium blockchain paradigm on hyperledger-based peer-to-peer lending system. China Commun 16(8):38–50. https://doi.org/10.23919/jcc.2019.08.004
45. Ali Syed T, Alzahrani A, Jan S, Siddiqui MS, Nadeem A, Alghamdi T (2019) A comparative analysis of blockchain architecture and its applications: problems and recommendations. IEEE Access 7:176838–176869. https://doi.org/10.1109/access.2019.2957660
46. Ma J, Lin S-Y, Chen X, Sun H-M, Chen Y-C, Wang H (2020) A blockchain-based application system for product anti-counterfeiting. IEEE Access 8:77642–77652. https://doi.org/10.1109/access.2020.2972026
47. Liang X, Du Y, Wang X, Zeng Y (2019) Design of a double-blockchain structured carbon emission trading scheme with reputation. In: 2019 34rd youth academic annual conference of Chinese association of automation (YAC). https://doi.org/10.1109/yac.2019.8787720

Printed in the United States
by Baker & Taylor Publisher Services